Commutative algebra is at the crossroads of algebra, number theory and algebraic geometry. This textbook, intended for advanced undergraduate or beginning graduate students with some previous experience of rings and fields, covers roughly the same material as Chapters 1–8 of Atiyah and Macdonald [A & M], but is cheaper, has more pictures, and is considerably more opinionated.

Alongside standard algebraic notions such as generators of modules and the ascending chain condition, the book develops in detail the geometric view of a commutative ring as the ring of functions on a space. The starting point is the Nullstellensatz, which provides a close link between the geometry of a variety V and the algebra of its coordinate ring $A = k[V]$; however, many of the geometric ideas arising from varieties apply also to fairly general rings.

The final chapter relates the material of the book to more advanced topics in commutative algebra and algebraic geometry. It includes an account of some famous "pathological" examples of Akizuki and Nagata, and a brief but thought-provoking essay on the changing position of abstract algebra in today's world.

T0282311

LONDON MATHEMATICAL SOCIETY STUDENT TEXTS

Managing editor: Dr C.M. Series, Mathematics Institute
University of Warwick, Coventry CV4 7AL, United Kingdom

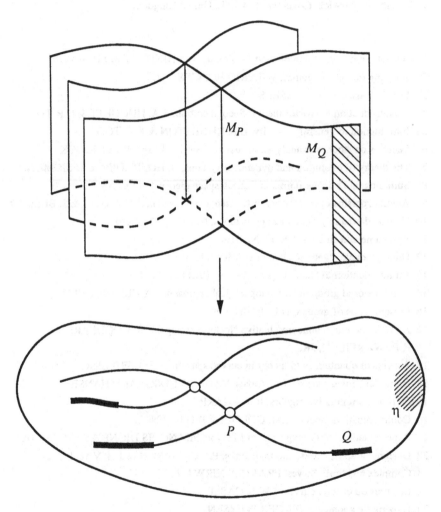

Frontispiece: let A be a ring and M an A-module ...

London Mathematical Society Student Texts 29

Undergraduate Commutative Algebra

Miles Reid
University of Warwick

CAMBRIDGE
UNIVERSITY PRESS

Published by the Press Syndicate of the University of Cambridge
The Pitt Building, Trumpington Street, Cambridge CB2 1RP
40 West 20th Street, New York, NY 10011-4211, USA
10 Stamford Road, Oakleigh, Melbourne 3166, Australia

© Cambridge University Press 1995

First published 1995

A catalogue record for this book is available from the British Library

Library of Congress cataloging in publication data

Reid, Miles (Miles A.)
Undergraduate commutative algebra / Miles Reid.
   p.   cm. - (London Mathematical Society student texts; 29)
Includes bibliographical references
ISBN 0 521 45255 4. - ISBN 0 521 45889 7 (pbk.)
1. Commutative algebra. I. Title. II. Series.
QA251.3.R45   1995
512'.24--dc20   94-27644   CIP

ISBN   0 521 45255 4   hardback
ISBN   0 521 45889 7   paperback

Transferred to digital printing 2002

Contents

Illustrations

Preface

These are notes from a commutative algebra course taught at the University of Warwick several times since 1978. In addition to standard material, the book contrasts the methods and ideology of abstract algebra as practiced in the 20th century with its concrete applications in algebraic geometry and algebraic number theory.

0
Hello!

This chapter contains a preliminary discussion of the aims and philosophy of the book, and is not logically part of the course. Some of the material may be harder to follow here than when it is treated more formally later in the book, so if you get stuck on something, don't worry too much, just skip on to the next item.

0.1 Where we're going

The purpose of this course is to build one of the bridges between algebra and geometry. Not the Erlangen program (linking geometries via transformation groups with abstract group theory) but a quite different bridge linking rings A and geometric objects X; the basic idea is that it is often possible to view a ring A as a certain ring of functions on a space X, to recover X as the set of maximal or prime ideals of A, and to derive pleasure and profit from the two-way traffic between the different worlds on each side.

Algebra here means rings, always commutative with a 1, and usually closely related to a polynomial ring $k[x_1, \dots, x_n]$ or $\mathbb{Z}[x_1, \dots, x_n]$ over a field k or the integers $\mathbb{Z}$, or a ring obtained from one of these by taking a quotient by an ideal, a ring of fractions, a power series completion, and so on; also their ideals and modules. In this book, A usually stands for a ring and k for a field, and I sometimes use these notations without comment. We're interested in questions such as zerodivisors (that is, $0 \neq x, y \in A$ such that $xy = 0 \in A$), factorisation (that is, writing $a \in A$ as a product $a = b \prod p_i^{n_i}$ with b invertible and p_i prime elements), similar questions for ideals, prime ideals, extension rings $A \subset B$, etc.

The study of rings of this type includes most of algebraic number theory and a large fraction of algebraic geometry. The methods for

1

studying them, by and large, are either simple algebraic arguments, or depend on the link with geometry which I want to introduce here. Thus, for example, rings have a *dimension theory*, in which $\dim k[x_1, \ldots, x_n] = n$ and $\dim \mathbb{Z}[x_1, \ldots, x_n] = n + 1$ (yes, $n + 1$ is right!), and already the language suggests a cocktail of two different subjects. The same holds for *local ring*, an idea at the very heart of commutative algebra.

0.2 Some definitions

Before describing briefly the geometric side and some aspects of the bridge, I recall a few very elementary algebraic topics and introduce some definitions, which I hope are mostly already familiar.

Let A be a ring, commutative with a 1. *Zerodivisors* of A are nonzero elements $x, y \in A$ such that $xy = 0$. If A has no zerodivisors and $A \neq 0$, it is an *integral domain*; note that $0 \neq 1$ is part of the definition of integral domain. An integral domain is contained in a unique field K such that every element of K is a fraction a/b with $a, b \in A$ and $b \neq 0$; this is the *field of fractions* of A, sometimes written $K = \operatorname{Frac} A$, and I assume you understand its construction. An element $x \in A$ is *invertible* or a *unit* of A if it has an inverse in A, that is, there exists $y \in A$ such that $xy = 1$.

An element $x \in A$ is *nilpotent* if $x^n = 0$ for some n. Prove for yourself that x nilpotent implies that $1 - x$ is invertible in A. [Hint: write out $(1 - x)^{-1}$ as a power series.] Prove also that x and y nilpotent implies that $ax + by$ is nilpotent for all $a, b \in A$, so that the set of nilpotent elements of A is an ideal, the *nilradical* nilrad A. [Hint: use the binomial theorem.] An element $x \in A$ is *idempotent* if $x^2 = x$. Obviously if x is idempotent then so is $x' = 1 - x$, and then $x + x' = 1$ and $xx' = 0$ (please check all this for yourself), so that x and x' are *complementary orthogonal idempotents*; now by writing $a = ax + ax'$ for any $a \in A$, you see that A is a direct sum of rings $A = A_1 \oplus A_2$, where $A_1 = Ax$ and $A_2 = Ax'$.

0.3 The elementary theory of factorisation

Suppose that A is an integral domain. A nonzero element $x \in A$ is *irreducible* if x itself is not invertible, and $x = yz$ with $y, z \in A$ implies that either y or z is invertible. $x \in A$ is a *prime element* if it is not a unit, and $x \mid yz$ implies either $x \mid y$ or $x \mid z$. It is trivial to see that prime implies irreducible, but the other way round is false in general.

A is a *UFD* (*unique factorisation domain*) if (i) every element x factors as a product of finitely many irreducibles $x = \prod x_i$ with x_i irreducible, and (ii) irreducible implies prime.

Proposition *In a UFD A, the expression of $x = b \prod p_i^{n_i}$ as a product of irreducibles with $p_i \nmid p_j$ is unique (up to invertible elements).*

I assume you know this. Otherwise, see any textbook on algebra, for example, [C], [H & H], Chapter 4 or [W]. In the following sections, I need to assume that the polynomial ring $k[x_1, \ldots, x_n]$ is a UFD; the proof of this is discussed in Exs. 0.8–9 below.

0.4 A first view of the bridge

For simplicity, and to be able to describe in a few intuitive words a representative case of the geometric side, suppose that k is an algebraically closed field, for example $k = \mathbb{C}$. Then the polynomial ring $k[x_1, \ldots, x_n]$ is a ring of functions on k^n, because a polynomial $g \in k[x_1, \ldots, x_n]$ is a function $g = g(x_1, \ldots, x_n)$ of $x_1, \ldots, x_n$. Moreover, evaluating a polynomial at a point $P = (a_1, \ldots, a_n) \in k^n$ determines a homomorphism $k[x_1, \ldots, x_n] \to k$ defined by $g \mapsto g(P)$, whose kernel is the maximal ideal $m_P = (x_1 - a_1, \ldots, x_n - a_n)$ (see Ex. 1.15). This is the correspondence between a ring A and a space X in ideal form: $A = k[x_1, \ldots, x_n]$ is the ring of polynomial functions on $X = k^n$, and the points of X correspond to maximal ideals of A.

0.5 The geometric side – the case of a hypersurface

Suppose that $0 \neq F \in k[x_1, \ldots, x_n]$; then the locus

$$X = V(F) = \{P = (a_1, \ldots, a_n) \mid F(P) = 0\} \subset k^n$$

is a hypersurface. It is $(n-1)$-dimensional because you can (almost always) use the equation $F = 0$ to solve for a_1 in terms of $a_2, \ldots, a_n$.

Now consider the quotient ring $A = k[x_1, \ldots, x_n]/(F)$, that is, the ring of residue classes modulo the ideal generated by F. Then an element $g \in A$ defines a k-valued function on X: indeed, if g is the class in A of a polynomial $\tilde{g} \in k[x_1, \ldots, x_n]$ then for $x \in X$ the value $g(x) = \tilde{g}(x)$ does not depend on the choice of $\tilde{g}$.

Now we can see (fairly light-weight) traffic passing over the bridge. First, to what extent can A be viewed as a ring of functions on X?

(i) If F has no multiple factors, say $F = \prod f_i$ with $f_i \nmid f_j$, then it can be shown that F generates the ideal of all functions vanishing on X. (You can try the proof as an exercise after Chapter 5, see Ex. 5.5.) It follows from this that an element $g \in A$ is uniquely determined by the corresponding function $g \colon X \to k$, so that A is contained in the ring of k-valued functions on X.

On the other hand, if F has a multiple factor, say $F = f^k g$ with $k \geq 2$ then also $fg = 0$ everywhere on X, and hence F does not generate the ideal of all functions vanishing on X. At the same time, A has nonzero nilpotent elements (because $x = \operatorname{im} fg \in A$ satisfies $x^k = 0$). In this case, it is not reasonable to try to view the nilpotent element x as a function on X, because it is zero everywhere on X. Thus

F has a multiple factor

$\Longleftrightarrow$ more functions vanish on X than (F)

$\Longleftrightarrow$ A has nilpotent elements

$\Longleftrightarrow$ A has nonzero elements that are 0 as functions on X.

(ii) If F has a factorisation $F = f_1 f_2$, where f_1, f_2 are polynomials with no common factors, then A has zerodivisors (because $x_1 = \operatorname{im} f_1$, $x_2 = \operatorname{im} f_2$ satisfy $x_1 \neq 0, x_2 \neq 0$ but $x_1 x_2 = 0$); this corresponds to a decomposition $X = X_1 \cup X_2$ of X as a union of two smaller hypersurfaces X_i given by $f_i = 0$ for $i = 1, 2$. Thus

A has zerodivisors (not nilpotents)

$\Longleftrightarrow$ X is reducible: $X = X_1 \cup X_2$.

That is, something in algebra equals something in geometry.

(iii) I mentioned complementary orthogonal idempotents and direct sums of rings in 0.2; you can't get much more abstract algebraic than that. However, it is easy to see that A has nontrivial idempotents if and only if X is a disjoint union of two hypersurfaces, $X = X_1 \sqcup X_2$. If $k = \mathbb{C}$, this just means that $X \subset \mathbb{C}^n$ is a disconnected topological space; you can't get much more geometric than that. The ring of functions (say, continuous) on a disconnected space $X = X_1 \sqcup X_2$ is a direct sum of the rings of functions on X_1 and X_2.

(iv) We will see that there is a close relation between ideals $I \subset A$ and subvarieties of X; we can already see that if $I \subset A$ is an ideal then it defines a *subvariety* $V(I) \subset X$, the subset of $P \in X$ where

$f(P) = 0$ for all the functions $f \in I$. But this is quite a long story that I defer until later. For the time being, I state without proof the following result (a special case of the weak Nullstellensatz, see Theorem 4.10 and 5.1).

Proposition *Maximal ideals of A are in one-to-one correspondence with points $P \in X$. That is,*

$$P = (a_1, \ldots, a_n) \in X \; \leftrightarrow \; m_P = (x_1 - a_1, \ldots, x_n - a_n) \subset A.$$

To repeat my refrain, something in algebra (the maximal ideals of A) equals something in geometry (the points of X).

(v) Assume that $F \in k[x_1, \ldots, x_n]$ is irreducible, so that A is an integral domain. When is $A = k[x_1, \ldots, x_n]/(F)$ a UFD?

For example, if $F = xz - y^2 \in k[x, y, z]$ then $xz = y^2$ holds in the quotient ring A, whereas it is not hard to check that x, y, z are irreducible; therefore A is not a UFD. Now draw the picture of the locus $X : (xz = y^2)$, which is the ordinary quadric cone (see Figure 0.5). I will come back to this picture several times later in the book. Observe that X is a cone, and so contains lots of lines, for example, the lines $L_\lambda \subset X$ defined by $x = \lambda^2 z, y = \lambda z$; these are codimension 1 subvarieties of X.

I take $\lambda = 0$ to simplify the notation, so consider the line $L = L_0 \subset X$ defined by $x = y = 0$. The special feature of X is that the ideal $I_L \subset A$ of functions vanishing on L is not generated by one element. In fact $I_L = (x, y)$, but y also vanishes along a second line $y = z = 0$, whereas x vanishes along L with multiplicity 2. Geometrically, this corresponds to the fact that the plane $x = 0$ is everywhere tangent to X along L; or, to put it another way, at any point where $z \neq 0$, I have $x = y^2/z$. In this sense, L is not locally defined by one equation.

In other words, the geometric question of codimension 1 subvarieties and how to define them by equations is closely related to the algebraic question of unique factorisation in the ring A.

0.6 ℤ versus k[X]

The comparison between the ring of integers $\mathbb{Z}$ and the polynomial ring $k[X]$ in a single variable over a field k is one of the central points to be made at the outset in a commutative algebra course. From the algebraic

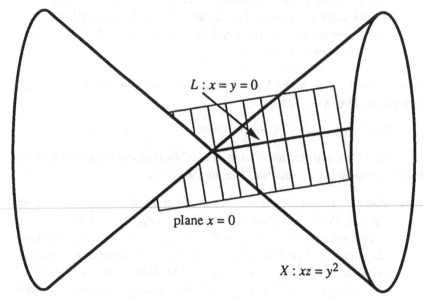

Figure 0.5. Quadric cone with a line

point of view, these two rings are very similar in many formal respects, and yet they are very different in substance. (Compare also Exs. 0.10–12 and the worked example 1.5.)

Points of similarity Recall that $\mathbb{Z}$ and $k[X]$ are both *Euclidean rings*, that is, integral domains satisfying "division with remainder" or the "Euclidean algorithm": for any a, b, there exists an expression $a = bq + r$ with r less than b. The ideal theory of the two rings proceeds in parallel from this fact: from division with remainder it follows easily that every ideal is *principal* (generated by one element), either (0) or (f) for an element f. I assume you know all this (see Exs. 0.1–9), including how you deduce the familiar $af + bg = h$ property of the highest common factor $h = \mathrm{hcf}(f, g)$, unique factorisation, etc.; if not, see, for example, [C] or [H & H], Chapter 4.

Points of difference Obviously $\mathbb{Z}$ and $k[X]$ are as different as chalk and cheese, but it is worthwhile trying to pin down the difference. I give two illustrations. First an algebraic statement: $k[X]$ contains a field, whereas $\mathbb{Z}$ doesn't. As you know, for any ring A, there is a unique homomorphism $t: \mathbb{Z} \to A$: after taking $1 \mapsto 1_A$, the rest is forced. If A contains a field then t factors through the prime subfield, either $\mathbb{F}_p$ or $\mathbb{Q}$. If a ring A contains a field k as a subring, then A and any A-module are k-vector spaces. In the case $A = \mathbb{Z}$, there is obviously no way of embedding either $\mathbb{F}_p$ or $\mathbb{Q}$ into A. The same holds for $A = \mathbb{Z}/(p^2)$, with p a prime number: the additive group $\mathbb{Z}/(p^2)$ is not a vector space over any field; see Ex. 0.10. For this reason, one sometimes says that a ring containing a field k has *equal characteristic* char k (either 0 or p), whereas a ring like $\mathbb{Z}$ has *unequal characteristic*, in the sense that the ring itself has characteristic 0, but it has residue fields $\mathbb{F}_p$ of characteristic p.

Here is another difference: $k[X]$ contains *variables*. To put it algebraically, a typical maximal ideal of $k[X]$ is (X), and it makes sense to *differentiate* with respect to X; that is, there is a k-linear map

$$\frac{\mathrm{d}}{\mathrm{d}X}: k[X] \to k[X], \qquad \text{defined by } x^n \mapsto nx^{n-1} \text{ for all } n,$$

with the properties and applications that you know about. By contrast, the maximal ideals of $\mathbb{Z}$ are (2), (3), (5), etc., and

$$\frac{\mathrm{d}}{\mathrm{d}2}, \frac{\mathrm{d}}{\mathrm{d}3}, \frac{\mathrm{d}}{\mathrm{d}5}, \dots$$

are of course completely meaningless. There is no nonzero derivation of $\mathbb{Z}$ to anything.

To put it another way, multiplication by a natural number n is the additive operation $a \mapsto a + \cdots + a$ (with n summands), and therefore this operation, and the ideal (n) generated by n, are already determined by the additive structure.

0.7 Examples

Recall that we intend to study extension rings of $\mathbb{Z}$ and of k, and that the distinctions in 0.6 will carry over to these. I continue the theme of 0.6 with two slightly more substantial examples from algebraic geometry and algebraic number theory illustrating this and other points.

Example 1 Suppose that k is an algebraically closed field of characteristic $\neq 2$, and let A be the ring $A = k[X,Y]/(Y^2 - X^3)$. By the correspondence of 0.5, A is a ring of functions on the plane curve $C \subset k^2$ given by $Y^2 = X^3$. See Figure 0.7(a).

Now A can be viewed as the extension ring $k[X][\sqrt{X^3}]$ obtained by adjoining the square root of X^3 to $k[X]$. If $(X - \alpha)$ is a maximal ideal of $k[X]$ with $\alpha \in k$, it is contained in maximal ideals $(X - \alpha, Y - \beta)$, corresponding to the square roots $\beta = \pm\sqrt{\alpha^3}$; there are obviously two of these if $\alpha \neq 0$, and one otherwise. These ideals require two generators. Moreover, it is easy to see that the elements X and Y are irreducible in A, but not prime: indeed, $Y^2 = X^3$ in A, so that $X \mid Y^2$, but $X \nmid Y$.

At this point, observe that I am doing something fairly silly by taking the square root of X^3. It is clearly much more sensible to take the square root of X instead. Let $A' = k[t]$ where $t = Y/X = \sqrt{X}$; this is a slightly bigger ring that A. Then in it, $X = t^2$ and $Y = t^3 \in A' = k[t]$, so that A' is just a polynomial ring, so of course it is a UFD, and every ideal is principal.

Example 2 Now consider $B = \mathbb{Z}[\sqrt{-3}]$, the extension ring obtained by adjoining $\sqrt{-3}$ to $\mathbb{Z}$. What are its maximal ideals? If P is a nonzero prime ideal of B then $P \cap \mathbb{Z}$ is a nonzero prime ideal of $\mathbb{Z}$, so of the form (p); we say that P *lies over* p. I check first that every prime number $p \neq 2, 3$ either splits as a product $p = f_+ f_-$ of two prime elements of B, or remains a prime element of B, and, in particular, any prime ideal of B not lying over 2 is principal. Indeed, any $p \equiv 1 \bmod 6$ can be written as $p = 3a^2 + b^2$ with $a, b \in \mathbb{Z}$; this was proved in 1760 by Euler, and is worked out in Ex. 0.14. Thus p factors in B as a product of two irreducible elements $p = f_+ f_-$, with $f_\pm = b \pm a\sqrt{-3}$; for example,

$$7 = (2 + \sqrt{-3})(2 - \sqrt{-3}).$$

It is easy to check that $B/(f_\pm) \cong \mathbb{F}_p$, so that $f_\pm$ are two prime elements of B lying over p. If $p \equiv 5 \bmod 6$ then $X^2 + 3$ is irreducible in $\mathbb{F}_p[X]$ by quadratic reciprocity (see Ex. 0.13), and $B/(p) \cong \mathbb{F}_p[X]/(X^2 + 3) \cong \mathbb{F}_{p^2}$ is a field, so that p is a prime element of B. It is also not hard to check that $(\sqrt{-3}) = 3\mathbb{Z} \oplus \mathbb{Z}\sqrt{-3} \subset B$, so that $B/(\sqrt{-3}) = \mathbb{F}_3$, and again $\sqrt{-3}$ is a prime element of B.

However, 2 is bad in B: you can easily prove it is irreducible, because

$$(a + b\sqrt{-3})(a - b\sqrt{-3}) = a^2 + 3b^2 \neq 2 \qquad \text{for any } a, b \in \mathbb{Z},$$

but it is not a prime element since $2^2 = (1 + \sqrt{-3})(1 - \sqrt{-3})$. Thus the prime ideal over 2 in B is $(2, 1 + \sqrt{-3})$, which needs 2 generators.

At this point, anyone who knows algebraic number theory will see that I am doing something fairly silly by taking the square root of -3. It is clearly much more sensible to take $\omega = (-1 + \sqrt{-3})/2$ instead; note that $\omega = \exp(2\pi i/3)$ is a primitive cube root of 1, satisfying $\omega^2 + \omega + 1 = 0$. Let $B' = \mathbb{Z}[\omega]$, a slightly bigger ring than B. The analysis of its prime

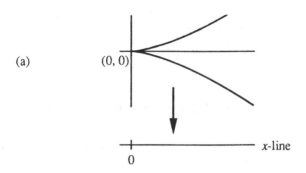

(a)

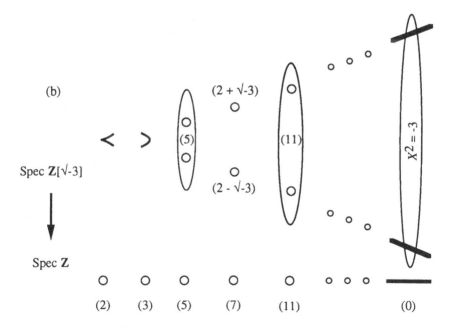

(b)

Spec **Z**[√-3]

Spec **Z**

Figure 0.7. The cuspidal cubic and Spec $\mathbb{Z}[\sqrt{-3}\,]$

ideals is exactly as for B except above 2, which is a prime element of B', because $B'/(2) = \mathbb{F}_2[\omega] = \mathbb{F}_4$, a quadratic extension field of $\mathbb{F}_2$. In fact you can prove that B' is a UFD (see Ex. 0.16).

Figure 0.7(b) draws the prime ideals of $B = \mathbb{Z}[\sqrt{-3}]$ in schematic form. I draw two points in a bubble over the primes $p \equiv 5 \bmod 6$ to represent the single prime $p \in B$, because I have in mind the two conjugate points $X = \pm\sqrt{-3}$ of the X-line defined over $\mathbb{F}_{p^2}$.

0.8 Reasons for studying commutative algebra

Commutative algebra is the crossroads between algebraic number theory, algebraic geometry and abstract algebra. Although much of the material of this book develops techniques of algebra, it should be clear that my main interest is the applications of these ideas to geometry and number theory.

a. Algebraic number theory Galois theory studies field extensions, with motivation coming from the study of polynomials and their roots; thus, corresponding to a polynomial

$$f(X) = a_n X^n + a_{n-1} X^{n-1} + \cdots + a_0 \in k[X]$$

with coefficients in a field k, one knows how to build a field extension $k \subset K$ over which f has a root, or splits into linear factors. It often happens that k contains a subring $A \subset k$ of interest, for example $\mathbb{Z} \subset \mathbb{Q}$, such that the coefficients of f are contained in A; we might then want to study a subring of K corresponding to A, for example the subring B generated by A and the roots of f. As a famous example, let $\varepsilon = \exp(2\pi i/n) = \sqrt[n]{1} \in \mathbb{C}$; over the ring $B = \mathbb{Z}[\varepsilon]$, a number of the form $x^n - z^n$ with x and z coprime factorises into n factors:

$$x^n - z^n = \prod_{i=0}^{n-1} (x - \varepsilon^i z).$$

Suppose we happened to know that $B = \mathbb{Z}[\varepsilon]$ is a UFD; then comparing two factorisations into primes coming from the left and right sides of $\prod_{i=0}^{n-1}(x - \varepsilon^i z) = y^n$ would obviously impose very strong restrictions on integer solutions of $x^n + y^n = z^n$, and in fact it is known that this can be used to prove that there are only the trivial solutions with x or y or $z = 0$; see, for example, [B & Sh], Chapter III, 1.1. Unfortunately, $\mathbb{Z}[\varepsilon]$

is not usually a UFD for large n, so you have to work harder if you want to prove this statement for all n and claim the due reward.

Rings like $\mathbb{Z}[\varepsilon]$ (or $B' = \mathbb{Z}[\omega]$ in 0.7) are called *rings of integers of number fields*, and the study of their ideals in the 19th century, in the context of Kummer's study of Fermat's last theorem (now Wiles' theorem), marks the start of commutative algebra.

b. Algebraic geometry Quite generally, an algebraic variety $V \subset k^n$ has a coordinate ring

$$k[V] = k[X_1, \ldots, X_n]/I_X = \{\text{polynomial functions } \varphi \colon V \to k\},$$

and the study of $k[V]$ gives lots of information about V. From the point of view of ring theory, every integral domain that is a finitely generated k-algebra occurs as $k[V]$ for some variety V.

For example, if you study a plane curve such as $C \colon (y^2 = x^3) \subset k^2$, you will rapidly come to the conclusion that its properties are closely related to those of the ring $A = k[X, Y]/(Y^2 - X^3)$. Thus, as we saw in 0.7, Example 1, A can be embedded in a bigger ring $A' = k[t]$ with $X = t^2, Y = t^3$, which corresponds to the parametrisation $x = t^2, y = t^3$ of the curve C. The fact that the origin $(0, 0)$ is a singular point of C is reflected in the algebra, as we saw in 0.7, and will see in several places below.

c. Abstract algebra Commutative rings were studied extensively at the turn of the century: as I have just sketched, they occurred in the 19th century as the rings of integers of number fields, and as the rings $k[V]$ associated with algebraic varieties $V \subset k^n$. In either case, these are quotients of polynomial rings $A = \mathbb{Z}[x_1, \ldots, x_n]/I$ or $A = k[x_1, \ldots, x_n]/I$; in the early 20th century, notably in the work of Hilbert, Emmy Noether, Krull, Emil Artin and others, they occurred as abstract structures satisfying the ring axioms, and whatever other axioms were needed to get reasonable results, notably the a.c.c. (ascending chain condition, see Chapter 3). A very important motivating idea for the development of algebra in the 1910s and 1920s was the fact that the abstract approach is often simpler and more general, and yields many of the results for the concrete quotients of polynomial rings with considerably less effort; see, for example, Remark 1.5. In this vein, this course discusses a number of results which are showcases of the methods of abstract algebra; at the same time, I point out some problems where the abstract approach has its limitations.

In conclusion, rings of the form $\mathbb{Z}[X_1, \ldots, X_n]/I$ or $k[X_1, \ldots, X_n]/I$ for suitable ideals I are important in both algebraic geometry and algebraic number theory, and have also been central to the development of algebra.

0.9 Discussion of contents

The course can be viewed as a continuation of an algebra course on rings and modules, as taught in the second or third year in many British universities. The book covers roughly the same material as Atiyah and Macdonald [A & M], Chaps. 1–8, but is cheaper, has more pictures, and is considerably more opinionated.

However, rather than talking about abstract algebra for its own sake, my main aim is to discuss and exploit the idea that a commutative ring A can be thought of as the ring of functions on a space $X = \operatorname{Spec} A$. Here are some points I will try to emphasise along these lines:

(1) *Prime spectrum* $\operatorname{Spec} A$ The attempt to re-establish the space X as the *spectrum* or *prime spectrum*, the set of prime ideals of A. In important cases, the prime ideals are controlled by maximal ideals.

(2) *Geometric rings and the Nullstellensatz* A ring that can be written as the coordinate ring $A = k[V]$ of a variety V is by definition a ring of functions on V, and the correspondence between geometry and algebra is especially close in this case.

(3) *Localisation* A ring of fractions $A[1/f]$ corresponds to restricting functions on X to the open subset $X_f = \{x \in X \mid f(x) \neq 0\}$.

(4) *Primary decomposition* A module can be pictured as a geometric object living over a subset of X (see the Frontispiece).

(5) *Integral extensions and normalisation*

(6) *Discrete valuation rings* Discrete valuation rings (DVRs) are the best kind of UFDs, having only one prime. The idea of a discrete valuation corresponds closely to the idea of measuring the power of a prime p dividing an integer, or the order of zeros and poles of a function on a normal algebraic variety or complex analytic space.

(7) *A Noetherian normal ring is an intersection of DVRs.* This result was somehow omitted from [A & M], but is very important. The statement and its proof is an exemplary chain of abstract algebraic reasoning. However, the result includes (i) the statement

that a meromorphic function on a complex manifold having no poles in codimension one is holomorphic, and (ii) the characterisation of algebraic integers among all algebraic numbers in terms of p-adic valuations.

(8) *Finiteness of normalisation* An algebraic result that provides the ring of integers of a number field, and the resolution of singularities of algebraic curves. At the same time, it gives a salutary reminder: finiteness of normalisation holds for practically all rings of importance in the world, but not for all Noetherian rings.

0.10 Who the book is for

Although this course will give the student intending to study algebraic geometry or complex analytic geometry an introduction to basic material in commutative algebra, an equally important objective is to make accessible to algebraists and number theorists the benefits of geometric intuition in studying commutative rings. However uncomfortably they fit into the framework of axioms and abstract arguments, the pictures in this book are (in my opinion) what commutative algebra is all about.

If you are the kind of student of algebra who suffers from vertigo when more than one or two logical steps above the axioms, you may convince yourself that all the material is entirely rigorous, and, if so desired, treat the geometric stuff as fanciful digressions. Or maybe my approach is just not for you; by all means study the excellent book of Atiyah and Macdonald [A & M] instead.

0.11 What you're supposed to know

As already mentioned, this course assumes some prior knowledge of fields and rings, for example, polynomial rings, the elementary theory of factorisation, division with remainder and its application to proving that $\mathbb{Z}$ and $k[X]$ are PIDs (principal ideal domains), therefore UFDs. This stuff is treated in any number of books, for example [C], [H & H], Chapter 4 or [W]. The first few (very easy!) exercises to this section sketch some standard material I will assume later, for example the fact that the polynomial rings $\mathbb{Z}[x_1, \ldots, x_n]$ and $k[x_1, \ldots, x_n]$ are UFDs. I also use some basic material in set theory, notably the definition of partially ordered set, but I give a discussion of Zorn's lemma and the a.c.c. from first principles.

This course is intended for advanced undergraduates or beginning graduate students, and in addition to background from algebra it assumes basic facts from analysis such as radius of convergence of a power series and the definition of a topological space. On the other hand, I develop modules from scratch and (except for Appendix 8.13 to Chapter 8) do not assume any material from Galois theory beyond the simple lemma on primitive field extensions sketched in Ex. 0.5.

The relation between commutative algebra, algebraic geometry and algebraic number theory is symbiotic, and the student will derive valuable motivation from an interest in one or other of these subjects. However, I have tried to keep this course reasonably self-contained, sometimes at the expense of repeating material that is covered perfectly well elsewhere.

Exercises to Chapter 0

Exercises 1–9 are intended to sketch out some material which you are supposed to know.

0.1   Let A be any ring, and consider the polynomial ring $A[T]$. Prove that T is not a zerodivisor in $A[T]$. Generalise the argument to prove that a monic polynomial

$$f = T^n + a_{n-1}T^{n-1} + \cdots + a_0$$

(a polynomial with leading coefficient 1) is not a zerodivisor in $A[T]$.

0.2   Let A be any ring, $a \in A$ and $f \in A[T]$. Prove that there exists an expression $f = (T - a)q + r$ with $q \in A[T]$ and $r \in A$. [Hint: subtract off a suitable multiple of $(T - a)$ to cancel the leading term, then use induction on $\deg f$.] By substituting $T = a$, show that $r = f(a)$. (This result is often called the *remainder theorem* in algebra textbooks.)

0.3   Remind yourself that (i) $\mathbb{Z}$ and $k[T]$ are Euclidean rings, that is, they have division with remainder (see [C], [H & H], Chapter 4 or [W]); (ii) a ring having division with remainder is a PID; and (iii) a PID is a UFD.

0.4   Let $A[T]$ be the polynomial ring over a ring A, and B any ring. Suppose that $\varphi: A \to B$ is a given ring homomorphism; show that ring homomorphisms $\psi: A[T] \to B$ extending φ are in one-to-one correspondence with elements of B.

0.5   Remind yourself that (i) a nonzero prime ideal $m \subset k[T]$ is of the form (f) with f an irreducible polynomial, and the quotient ring $k[T]/m$ is a finite algebraic extension field of k in which f

has a root; and (ii) if A is a ring containing a field k and $\alpha \in A$ any element then the subring $k[\alpha] \subset A$ generated by k and α is isomorphic to $k[T]$ or is a field extension of the form $k[T]/(f)$ with f irreducible.

0.6 Let $B = k[T]$ with k a field; a *k-automorphism* of B is a ring homomorphism $\varphi \colon B \to B$ that is the identity on k and is an automorphism of B (that is, one-to-one and onto). (i) Describe the group $\mathrm{Aut}_k B$ of k-automorphisms of B; (ii) now do the same for the field $K = k(T)$ of rational functions. [Hint: use Ex. 0.4. The answer to (ii) is in terms of 2×2 matrixes.]

0.7 Let A be a UFD, K its field of fractions and $f \in A[T]$ a monic polynomial (defined in Ex. 0.1 above). Prove that if f has a root $\alpha \in K$ then in fact $\alpha \in A$. [Hint: you can probably remember this for $\mathbb{Z} \subset \mathbb{Q}$; write $\alpha = p/q$ where p, q have no common factors, and consider the equation $f(p/q) = 0$.]

0.8 Let A be a UFD and K its field of fractions. A polynomial

$$g = b_n T^n + b_{n-1}T^{n-1} + \cdots + b_0 \in A[T]$$

is *primitive* if its coefficients b_i have no common factors in A (other than units). Prove Gauss' lemma: the product of two primitive polynomials is primitive.

A polynomial $f \in K[T]$ has a *reduced expression* $f = af_0$ where $a \in K$, and $f_0 \in A$ is primitive, unique up to multiplying by a unit. (This just amounts to clearing denominators and taking out any common factor.) The point of Gauss' lemma is that if $f = af_0$ and $g = bg_0$ are reduced expressions for f and g then $fg = (ab)(f_0g_0)$ is a reduced expression for fg.

0.9 Prove that A a UFD implies $A[T]$ a UFD. [Hint: you know that $K[T]$ is a UFD, where $K = \mathrm{Frac}\,A$. Use Gauss' lemma to compare factorisation in $K[T]$ and $A[T]$.] Deduce that the polynomial rings $\mathbb{Z}[x_1, \ldots, x_n]$ and $k[x_1, \ldots, x_n]$ are UFDs.

0.10 Let $k = \mathbb{F}_p \cong \mathbb{Z}/(p)$ be the finite field with p elements. Compare the two rings $k[T]/(T^n)$ and $\mathbb{Z}/(p^n)$ for $n \geq 2$. Show that their elements can be written respectively in the form

$$a_0 + a_1 T + \cdots + a_{n-1}T^{n-1} \quad \text{or} \quad a_0 + a_1 p + \cdots + a_{n-1}p^{n-1}$$

with $a_i \in \{0, 1, \ldots, p-1\}$ for $i = 0, \ldots, n-1$; determine the addition and multiplication of these power series in the two rings, and note that they differ only by the p-adic "carry".

0.11   Prove that the ring $\mathbb{Z}/(p^n)$ for $n \geq 2$ does not contain a field, and cannot be made into a vector space over $\mathbb{F}_p$ in a way compatible with its Abelian group structure.

0.12   Show that it makes sense to take infinite power series

$$a_0 + a_1 T + \cdots + a_n T^n + \cdots \quad \text{or} \quad a_0 + a_1 p + \cdots + a_n p^n + \cdots,$$

and add and multiply them according to the rules you worked out in Ex. 0.10. The ring of all such power series are respectively the *formal power series ring* in one variable $k[\![T]\!]$ and the *ring of p-adic integers* $\mathbb{Z}_p$.

0.13   Use quadratic reciprocity to prove that -3 is a quadratic residue mod p if and only if $p = 2$ or $p \equiv 1 \bmod 6$. Deduce that

$$\mathbb{Z}[\sqrt{-3}]/(p) \cong \begin{cases} \mathbb{F}_p \oplus \mathbb{F}_p & \text{if } p \equiv 1 \bmod 6, \\ \mathbb{F}_{p^2} & \text{if } p \equiv 5 \bmod 6. \end{cases}$$

0.14   Prove that a prime number $p \equiv 1 \bmod 6$ is of the form $3a^2 + b^2$ with $a, b \in \mathbb{Z}$. [Hint (due to Axel Thue [N], supplied to me by Alan Robinson): pick any $a \in \mathbb{Z}$ with $a^2 \equiv -3 \bmod p$ (this is possible by Ex. 0.13). Consider $x - ay$ as (x, y) run independently through the integers in $[0, \sqrt{p}]$; there are $> p$ pairs (x, y), hence they are not all distinct $\bmod p$. If $x_1 - ay_1 \equiv x_2 - ay_2$ then

$$(x_1 - x_2)^2 + 3(y_1 - y_2)^2 = Np \quad \text{with } N = 1, 2 \text{ or } 3.$$

Now $N = 2$ is impossible, and if $N = 3$ you can divide through.]

0.15   Prove that the ring of Gaussian integers $\mathbb{Z}[i]$, where $i^2 = -1$, has division with remainder with respect to the function "absolute value squared" $d(x + yi) = x^2 + y^2$. Deduce that it is a principal ideal domain, hence a UFD. [Hint: any complex number z is of the form $z = \alpha + \beta$, with *integral part* $\alpha \in \mathbb{Z}[i]$ and *smallest residue* β in the shaded square of Figure 0.15(a). Apply this to a fraction $(c + di)/(a + bi)$ to get division with remainder

$$c + di = (a + bi)\alpha + (a + bi)\beta.$$

The reason that the argument works is that the shaded square of Figure 0.15(a) is contained inside the unit disk.]

0.16   Prove that the ring $\mathbb{Z}[\omega]$ is a UFD, where $\omega^2 + \omega + 1 = 0$. [Hint: as in the previous question, prove that $\mathbb{Z}[\omega]$ has division with

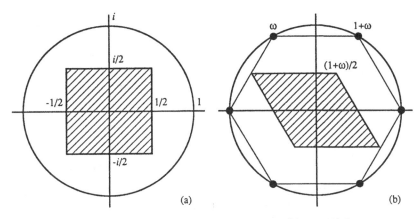

Figure 0.15. Smallest residue modulo $\mathbb{Z}[i]$ and $\mathbb{Z}[\omega]$

remainder with respect to the function "absolute value squared" $d(a+b\omega) = (a+b\omega)(a+b\omega^2) = a^2 - ab + b^2$, see Figure 0.15(b).]

0.17   Use the result of Ex. 0.16 to deduce another proof of the fact that any prime $p \equiv 1 \bmod 6$ is of the form $3a^2 + b^2$.

0.18   (Harder) Prove the cases $n = 3$ and $n = 4$ of Fermat's last theorem (see 0.8 and [B & Sh], Chapter III, 1.1).

0.19   Study the ring $B = \mathbb{Z}[\sqrt{5}]$ in the spirit of 0.7. A maximal ideal P of B lies over a prime $(p) = P \cap \mathbb{Z}$ of $\mathbb{Z}$. If $p \neq 2, 5$, then either (i) 5 is a quadratic residue mod p, and then p factors in B as a product of two elements $p = f_1 f_2$, and $B/(f_1) \cong B/(f_2) \cong \mathbb{F}_p$, so that f_1 are f_2 are prime elements of B (you can prove this by a minor modification of the method of Ex. 0.14); or (ii) 5 is a quadratic nonresidue mod p, and then $B/(p) \cong \mathbb{F}_p[\sqrt{5}]$ is a quadratic field extension of $\mathbb{F}_p$, so that p remains a prime element in B. (By quadratic reciprocity, the first happens if and only if $p \equiv \pm 1 \bmod 5$, for example, $11 = (4 - \sqrt{5})(4 + \sqrt{5})$.)

Also $\sqrt{5}$ is a prime element of B over 5, because $(\sqrt{5}) = \mathbb{Z}5 \oplus \mathbb{Z}\sqrt{5} \subset B$, giving $B/(\sqrt{5}) = \mathbb{F}_5$. However, 2 is bad in B: it is an irreducible element, because

$$(a\sqrt{5} + b)(a\sqrt{5} - b) = 5a^2 - b^2 \neq 2 \qquad \text{for any } a, b \in \mathbb{Z}$$

(argue mod 4), but not a prime element since

$$2^2 = (\sqrt{5}+1)(\sqrt{5}-1).$$

Thus the prime ideal over 2 in B is $(2, \sqrt{5}+1)$, which needs 2 generators.

0.20   Consider the ring $B' = \mathbb{Z}[\tau]$ where $\tau = (1+\sqrt{5})/2$ is the so-called "golden ratio", which satisfies $\tau^2 = \tau + 1$. Show that every maximal ideal of B' is principal. [Hint: compare 0.7.]

0.21   Show that an element $a+b\tau$ is a unit of B' (that is, is invertible in B') if and only if $N(a+b\tau) = a^2 + ab - b^2 = \pm 1$. Prove that the multiplicative group of units of B' is generated by $\pm 1, \tau$. [Hint: for the first part, note that

$$\frac{c+d\tau}{a+b\tau} = \frac{(c+d\tau)(a+b(1-\tau))}{(a+b\tau)(a+b(1-\tau))} = \frac{(c+d\tau)(a+b(1-\tau))}{a^2+ab-b^2}.$$

For the second part, if $a+b\tau$ is a unit then so is $\pm\tau(a+b\tau) = \pm(b+(a+b)\tau)$ and $\pm(\tau-1)(a+b\tau) = \pm(b-a+a\tau)$; one of these operations allows you to reduce $\max\{|a|,|b|\}$ until you get down to 1.]

0.22   Prove that $\mathbb{Z}[\tau]$ is a UFD. [Hint (provided by Jack Button): write $N(a+b\tau) = a^2+ab-b^2$. The aim is to show that $\mathbb{Z}[\tau]$ satisfies division with remainder with respect to the function $d(a+b\tau) = |N(a+b\tau)| = |a^2+ab-b^2|$. As in Exs. 0.15–16, every element of $\mathbb{Q}[\tau]$ has a *smallest residue* $r+s\tau$ mod $\mathbb{Z}[\tau]$ in a "shaded" square, with $r,s \in [-1/2, 1/2]$, and then obviously $|N(r+s\tau)| \le 3/4$.]

0.23   (Harder) Let $f \in A$; if f is reducible then the principal ideal (f) is contained in a bigger principal ideal (f_1). Consider the following conditions on a ring A:

(a) A is a UFD;
(b) every increasing chain $(f_1) \subset (f_2) \subset \cdots \subset (f_n) \subset \cdots$ of principal ideals is eventually stationary, that is $(f_k) = (f_{k+1}) = \cdots$ for some k;
(c) any element can be written as a product of irreducible elements.

Prove that (i) $\implies$ (ii) and (iii), and that (ii) $\implies$ (iii). Find counterexamples to the implications (iii) $\implies$ (i) and (ii) $\implies$ (i). Find a ring not satisfying (iii). Find a counterexample to the implication (iii) $\implies$ (ii).

1
Basics

1.1 Convention

I assume that the student already knows the definition of ring and ring homomorphism (see for example [A & M], p. 1); however, throughout this course, a ring A is also assumed to be commutative (that is, $ab = ba$) and to have an identity element $1 = 1_A \in A$; a ring homomorphism $\varphi \colon A \to B$ is assumed to satisfy $\varphi(1_A) = 1_B \in B$, and a subring of a ring is assumed to share the same identity element. I will use this convention from now on, usually without mention. See also the discussion in 1.11.

1.2 Ideals

By definition, an *ideal* of a ring A is a subset $I \subset A$ such that $0 \in I$, and

$$af + bg \in I \quad \text{for all } a, b \in A, \text{ and } f, g \in I.$$

That is, I can form linear combinations of elements of I with coefficients in A, and the result is still in I; I write (f_1, f_2) or $Af_1 + Af_2$ for the ideal generated by elements f_1, $f_2 \in A$, and analogous notation such as $(\Sigma) + fA + J$ for the ideal generated by a set Σ, an element f and an ideal J. Notice that 0 is an ideal (I write 0 or $\{0\}$ or (0) indiscriminately), as is A itself, and

$$1 \in I \iff I = (1) = A.$$

The following result explains the definition and main function of ideals, and is assumed known:

Proposition

(i) *The kernel of a ring homomorphism $\varphi \colon A \to B$ is an ideal.*

(ii) *If $I \subset A$ is an ideal then there exists a ring A/I and a surjective homomorphism $\varphi \colon A \to A/I$ such that $\ker \varphi = I$; the pair A/I and φ is uniquely defined up to isomorphism. The standard map* $\varphi \colon A \to A/I$ *is called the* quotient homomorphism.

(iii) *In the notation of (ii), the map*

$$\varphi^{-1} \colon \left\{ ideals\ of\ A/I \right\} \to \left\{ ideals\ of\ A\ containing\ I \right\}$$

is a one-to-one correspondence.

1.3 Prime and maximal ideals, the definition of $\operatorname{Spec} A$

Definition *$S \subset A$ is a* multiplicative set *if*

$$1 \in S \quad \text{and} \quad f, g \in S \implies fg \in S.$$

An ideal $P \subset A$ is *prime* if its complement $A \setminus P$ is a multiplicative set. Thus a prime ideal $P \subset A$ is an ideal such that $P \neq A$ (because I insist that $1 \in S$, so that $1 \notin P$), and $fg \in P$ implies $f \in P$ or $g \in P$; a ring A is an *integral domain* (or *integral ring*, or *domain*) if $0 \subset A$ is a prime ideal.

An ideal $m \subset A$ is *maximal* if $m \neq A$, but there is no ideal I strictly between m and A, in other words,

$$\left. \begin{array}{c} m \subset I \subset A \\ \text{with } I \text{ an ideal} \end{array} \right\} \implies m = I \text{ or } I = A.$$

Vacuous remark Having $1 \notin P$ and $0 \neq 1$ as part of the definition of a prime ideal and an integral domain is a convenient way of sidestepping awkward anomalies. You have probably seen this kind of question before, in the form: do you allow 1 as a prime number? I don't.

A ring A has distinguished elements 0_A and 1_A. It is convenient in lots of arguments to allow $A = (1_A)$ as an ideal. For this reason, the wise allow the possibility $0_A = 1_A$, but then of course $A = \{0\}$, with the addition and multiplication tables you can work out. Content-free considerations of this nature are referred to collectively as *empty set theory*. Compare Ex. 1.20.

The following result is easy, and I assume it known:

Proposition

(i) *P is prime $\iff A/P$ is an integral domain;*

(ii) *m is maximal* $\iff A/m$ *is a field.*

For any prime ideal P, the field of fractions $k(P) = \mathrm{Frac}(A/P)$ of the integral domain A/P is called the *residue field* at P. Thus every prime ideal P is obtained as the kernel of a homomorphism $A \to K$ of A to a field K, namely, the composite $A \twoheadrightarrow A/P \hookrightarrow k(P) = \mathrm{Frac}(A/P)$.

The *prime spectrum* or *Spec* of a ring A is the set of prime ideals of A, that is,

$$\mathrm{Spec}\, A = \{ P \mid P \subset A \text{ is a prime ideal} \}.$$

The *maximal spectrum* m-Spec A is the set of maximal ideals of A.

1.4 Easy examples

(1) A ring k is a field if and only if $0 \subset k$ is maximal (think about it!).

(2) The prime ideals of $\mathbb{Z}$ are familiar: $\mathrm{Spec}\,\mathbb{Z} = \{0, (2), (3), (5), \dots\}$.

(3) If k is a field and $A = k[X]$ then a prime ideal of A is 0 or (f) for f an irreducible polynomial. That is,

$$\mathrm{Spec}\, k[X] = \{0\} \cup \{ (f) \mid f \text{ is an irreducible polynomial} \}.$$

If k is algebraically closed, then the only irreducible polynomials are of the form $(\text{unit}) \cdot (X - \alpha)$, with $\alpha \in k$, so that prime ideals are 0 and $(X - \alpha)$ with $\alpha \in k$. That is, for k algebraically closed,

$$\mathrm{Spec}\, A = \{0\} \cup k.$$

(4) In the polynomial ring $k[X, Y]$ in 2 variables,

$$0 \subset (X) \subset (X, Y)$$

are all prime ideals, but only the last is maximal. More generally, if $(a, b) \in k^2$ then $m_{(a,b)} = (X - a, Y - b)$ is a maximal ideal of $k[X, Y]$, and the quotient homomorphism $k[X, Y] \to k[X, Y]/m_{(a,b)} = k$ is the map $f(X, Y) \mapsto f(a, b)$ "evaluate at (a, b)" (see Ex. 1.15). Thus as discussed in 0.5, (iv), viewing $k[X, Y]$ as the ring of functions on the (X, Y)-plane k^2 gives an interpretation of algebra (some of the maximal ideals of $k[X, Y]$) in terms of geometry (points of k^2).

(5) Similarly, if p is a prime number, then 0, (p), (X) and (p, X) are all prime ideals of $\mathbb{Z}[X]$, but only the last is maximal.

1.5 Worked examples: Spec $k[X, Y]$ and Spec $\mathbb{Z}[X]$

I now consider Spec A in the more complicated cases (4) $A = k[X, Y]$ and (5) $\mathbb{Z}[X]$ from the end of 1.4. At the same time as getting practice at working with prime ideals, I take the opportunity to rub in the message of 0.6, the analogy between $k[X]$ and $\mathbb{Z}$; although these rings are quite different in substance, the algebraic treatment of them is entirely parallel. I change the notation $\mathbb{Z}[X] \mapsto \mathbb{Z}[Y]$ to avoid confusing myself in the analogy.

Proposition *The prime ideals of $k[X, Y]$ are as follows:*

$$0; \ (f) \text{ for irreducible } f(X, Y) \in k[X, Y]; \text{ and maximal ideals } m.$$

Moreover, each maximal ideal is of the form $m = (p, g)$ where $p = p(X) \in k[X]$ is an irreducible polynomial in X (not a unit), and $g \in k[X, Y]$ a polynomial whose reduction modulo p is an irreducible element $\bar{g} \in (k[X]/(p))[Y]$; therefore the quotient $k[X, Y]/m$ is a finite algebraic extension field of k.

The prime ideals of $\mathbb{Z}[Y]$ are as follows:

$$0; \ (f) \text{ for irreducible } f \in \mathbb{Z}[Y]; \text{ and maximal ideals } m.$$

Moreover, each maximal ideal is of the form $m = (p, g)$ where p is a prime number and $g \in \mathbb{Z}[Y]$ a polynomial whose reduction modulo p is an irreducible element $\bar{g} \in \mathbb{F}_p[Y]$; therefore the quotient $\mathbb{Z}[Y]/m$ is a finite algebraic extension field of $\mathbb{F}_p$.

Proof Write $B = k[X]$ and $K = k(X)$ in the first case, and $B = \mathbb{Z}$ and $K = \mathbb{Q}$ in the second. Then in either case, B is a PID and K its field of fractions, and the ring under study is $A = B[Y]$. If a prime ideal P of $B[Y]$ is 0 or principal then there is nothing to prove, so that I can assume that P contains two elements f_1, f_2 with no common factor in $B[Y]$.

Step 1 I claim that f_1, f_2 also have no common factor in $K[Y]$.

This is an exercise in clearing denominators using Gauss' lemma (see Ex. 0.8). By contradiction, suppose that $f_1 = hg_1$ and $f_2 = hg_2$ with $h, g_1, g_2 \in K[Y]$ and $\deg h \geq 1$. As in Ex. 0.8, there exist reduced expressions $h = ah_0$, $g_1 = b_1\gamma_1$ and $g_2 = b_2\gamma_2$ with $a, b_1, b_2 \in K$, and h_0, γ_1, γ_2 primitive elements of $B[Y]$ (recall that this means their coefficients have no common factors). Now by Gauss' lemma $h_0\gamma_1$ and $h_0\gamma_2$

are again primitive, so that $f_1 = hg_1 = (ab_1)(h_0\gamma_1) \in B[Y]$ implies that $ab_1 \in B$, and similarly $ab_2 \in B$. Therefore $h_0 \mid f_1, f_2$, a contradiction.

Step 2 The ideal of A generated by f_1 and f_2 has nonzero intersection with B, that is, $(f_1, f_2) \cap B \neq 0$.

Indeed, $K[Y]$ is a PID, and $\mathrm{hcf}(f_1, f_2) = 1$, and therefore there exist $a, b \in K[Y]$ such that $af_1 + bf_2 = 1$. If $c \in B$ is a common denominator of all the coefficients of a and b then $B \ni (ca)f_1 + (cb)f_2 = c$.

Note that the arguments so far have only used the fact that B is a UFD.

Step 3 If P is a prime ideal of $A = B[Y]$ then $B \cap P$ is a prime ideal of B. We have seen in Steps 1–2 that if P is not principal then $B \cap P \neq 0$. Now B is a PID, so that any nonzero prime ideal is maximal. The proposition follows from this.

Remark You can read through the above proof for $k[X] \subset k(X)$ only, or for $\mathbb{Z} \subset \mathbb{Q}$ only, or more abstractly, thinking of $B \subset K$ as a general PID in its ring of fractions. The abstract algebraic approach is of course easier and more general, and applies directly to the concrete cases. This gives you a good feel for the attraction of the subject.

1.6 The geometric interpretation

The maximal ideals of $k[X, Y]$ have a geometric meaning: first think of the case of an algebraically closed field k. Then a maximal ideal is of the form $(X - a, Y - b)$, corresponding to a point $X = a, Y = b$ of the plane k^2.

In the general case, $k[X, Y]/m = F$ is an algebraic extension field of k. The quotient homomorphism $k[X, Y] \to F$ maps $X \mapsto a$, $Y \mapsto b$, and

$$m = k[X, Y] \cap (X - a, Y - b) \subset F[X, Y],$$

where $(X - a, Y - b)$ is the maximal ideal of $F[X, Y]$. In other words, since k is not algebraically closed, if you want a geometric interpretation of this maximal ideal as a point of the (X, Y)-plane, you must allow X, Y to take values in the extension field F, in the same way that pairs of complex conjugate roots help to understand roots of real polynomials.

In this case, Steps 1–2 of the proof consist of *eliminating* Y from the two equations $f_1(X, Y) = f_2(X, Y) = 0$ to get a single equation $p(X)$

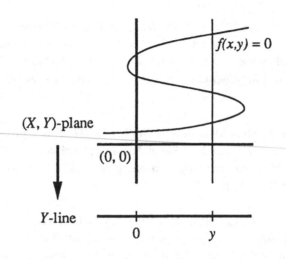

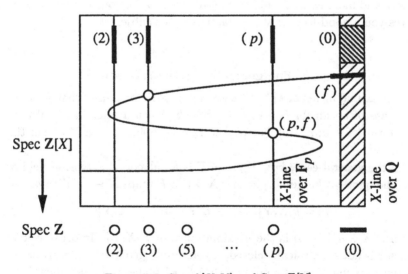

Figure 1.6. Spec $k[X, Y]$ and Spec $\mathbb{Z}[Y]$

for X: because $p \in (f_1, f_2)$ it follows that $f_1(X, Y) = f_2(X, Y) = 0 \implies p(X) = 0$.

The prime ideals (f) correspond to the irreducible curves $(f = 0)$ in k^2, and 0 to the whole plane k^2. See Figure 1.6(a).

The maximal ideal $m = (p, g)$ of $\operatorname{Spec} \mathbb{Z}[Y]$ corresponds in the same way to the point $\overline{g}(Y) = 0$ of the Y-line over the field $\mathbb{F}_p$. Since $\mathbb{F}_p$ is not algebraically closed, we must take roots of $\overline{g}(Y) = 0$ in an extension field of $\mathbb{F}_p$. In more detail, the field $F = \mathbb{F}_p[Y]/(\overline{g})$ is an algebraic extension of $\mathbb{F}_p$ in which $\overline{g}$ has a root $Y = \alpha$, and the quotient homomorphism $\mathbb{Z}[Y] \to F$ is given by $f \mapsto \overline{f} \mapsto \overline{f}(\alpha)$.

I am now going to draw $\operatorname{Spec} \mathbb{Z}[Y]$ schematically as Figure 1.6(b), although I know it is hard to follow when you first see it – don't worry too much about it. $\operatorname{Spec} \mathbb{Z}[Y]$ is pictured by analogy with $\operatorname{Spec} k[X, Y]$ as a kind of surface, the union of the Y-lines over $\mathbb{F}_p$ for each p.

The prime ideals (f) with $f \in \mathbb{Z}[Y]$ are of two kinds: either $f = p$ is a prime number, or $f = f(Y)$ has deg ≥ 1 and remains irreducible in $\mathbb{Q}[Y]$, thus defining a field extension $L = \mathbb{Q}[Y]/(f)$, that is, a point of the Y-line over $\mathbb{Q}$. I add (p) to the picture as the "curve" formed by the Y-line over $\mathbb{F}_p$, and (f) as the curve passing through all the points $\overline{f}(Y) = 0$ in the fibre over each p.

1.7 Zorn's lemma

The next two sections are concerned with the existence of prime ideals – recall that an overall aim is to discuss a fairly general ring A as a ring of functions on $\operatorname{Spec} A$, so I must at least show that A has lots of prime ideals.

This section is a preliminary digression in set theory. Many proofs in algebra involve a kind of infinite induction called *transfinite induction* that uses the axiom of choice, one of the axioms of set theory. One form of this axiom popular with algebraists is *Zorn's lemma*; as we will see below, it is very convenient to use. I treat this here because it is not properly taught in some undergraduate courses (for example, at Warwick in recent years).

Let Σ be a partially ordered set. Given a subset $S \subset \Sigma$, an *upper bound* of S is an element $u \in \Sigma$ such that $s < u$ for all $s \in S$. A *maximal element* of Σ is an element $m \in \Sigma$ such that $m < s$ does not hold for any $s \in \Sigma$. A subset $S \subset \Sigma$ is *totally ordered* if for every pair $s_1, s_2 \in S$ of elements of S, either $s_1 \leq s_2$ or $s_2 \leq s_1$.

Axiom (**Zorn's lemma**) *Suppose that* Σ *is a nonempty set with a partial order* $<$, *and that any totally ordered subset* $S \subset \Sigma$ *has an upper bound in* Σ. *Then* Σ *has a maximal element.*

The idea of transfinite induction is as follows: if $s \in \Sigma$ is not itself maximal, then I can choose s' such that $s < s'$, and so on, to get

$$s < s' < \cdots < s^{(\lambda)} < \cdots$$

This is a totally ordered subset, so has an upper bound in Σ; if the upper bound of the $s^{(\lambda)}$ is still not maximal, I can extend the totally ordered set some more. This cannot prove the axiom, because I need to make new choices at each stage (infinitely often). That is, there is never anything to stop me making finitely many choices, but logically speaking, I need a new axiom to be able to make infinitely many. It is proved in set theory that Zorn's lemma is equivalent to the axiom of choice (see Ex. 1.19).

1.8 Existence of maximal ideals

Proposition *Let A be a ring and $I \ne A$ an ideal; then there exists a maximal ideal m of A containing I.*

The proof of the proposition consists essentially of saying that if I is not already maximal, then it is contained in a bigger ideal, and so on, ... Zorn's lemma is needed to make "and so on, ... " into a rigorous transfinite induction.

Proof This is an easy illustration of how Zorn's lemma is used. Let Σ be the set of ideals $J \ne A$ containing I, ordered by inclusion $J_1 \subset J_2$. Then Σ is nonempty, since $I \in \Sigma$. If $\{J_\lambda\}$ for $\lambda \in \Lambda$ is a totally ordered subset of Σ then it is easy to see that $J^* = \bigcup J_\lambda \in \Sigma$ (that is, it is an ideal of A with $J^* \ne A$), and J^* is clearly an upper bound of $\{J_\lambda\}$. Thus I have verified that Σ satisfies all the assumptions of Zorn's lemma, so that it has a maximal element.   Q.E.D.

An element $a \in A$ of a nonzero ring A is *invertible*, or is a *unit* if $ab = 1$ for some $b \in A$, that is, if $(a) = A$. It is traditional to write $A^\times = \{\text{units of } A\}$ for the set of units, which is of course a group under multiplication, the *multiplicative group* of A (one also often sees A^*, but the star or asterix has many other uses).

Corollary $A = A^\times \sqcup \bigcup m$ *(disjoint union); that is, an element $f \in A$ is either a unit, or is contained in a maximal ideal, and not both.*

Proof If $a \in m$ for some maximal ideal m then, of course, a is not a unit; conversely, if a is not a unit then $(a) \neq A$, so that by the proposition, it is contained in a maximal ideal. Q.E.D.

1.9 Plenty of prime ideals

The previous proposition can be strengthened as follows, to give the existence of prime ideals.

Proposition *Let A be a ring, S a multiplicative set, and I an ideal of A disjoint from S; then there exists a prime ideal P of A containing I and disjoint from S.*

Proof Suppose that I can find an ideal $P \supset I$ maximal subject to the condition $P \cap S = \emptyset$. I claim that P is a prime ideal. For if $f, g \notin P$ then each of the ideals $P + Af$ and $P + Ag$ is strictly bigger than P, and therefore intersects S. Suppose that $p + af, q + bg \in S$ for $p, q \in P$ and $a, b \in A$; then, using the fact that S is multiplicative,

$$S \ni (p + af)(q + bg) = pq + pbg + qaf + abfg = p' + abfg,$$

with $p' \in P$. Now since $P \cap S = \emptyset$, it follows that $fg \notin P$.

Therefore, it is enough to find an ideal $P \supset I$ with $P \cap S = \emptyset$ and maximal subject to this condition. This is proved using Zorn's lemma in the same way as Proposition 1.8. Let Σ be the set of ideals J of A containing I and disjoint from S. Repeating the proof of Proposition 1.8 word-for-word gives that Σ has a maximal element. Q.E.D.

1.10 Nilpotents and the nilradical

You already know that $a \in A$ is *nilpotent* if $a^n = 0$ for some $n > 0$, and that the set of all nilpotent elements of A is an ideal of A, the *nilradical* nilrad A (compare 0.2). A ring A is *reduced* if nilrad $A = 0$, that is, A has no nonzero nilpotents. Proposition 1.9 implies the following result.

Corollary nilrad A *is the intersection of all prime ideals P of A, that is,*

$$\text{nilrad}\, A = \bigcap_{P \in \text{Spec}\, A} P. \tag{$*$}$$

In other words, $f \in A$ is not nilpotent if and only if there is a prime ideal $P \in \text{Spec}\, A$ such that $f \notin P$.

Proof If f is nilpotent then it obviously belongs to every prime ideal P. Conversely, suppose that f is not nilpotent; I claim that there exists a prime ideal $P \not\ni f$. For this, consider the multiplicative set generated by f, that is,

$$S = \{1, f, f^2, \dots\}.$$

Then $0 \notin S$ because f is not nilpotent. I apply Proposition 1.9 to this S and $I = 0$, and get a prime ideal P such that $P \cap S = \emptyset$, so $f \notin P$. Q.E.D.

1.11 Discussion of zerodivisors

An *idempotent* is an element $e \in A$ for which $e^2 = e$; it is easy to see (compare 0.2) that A has an idempotent $e \neq 0, 1$ if and only if it is a direct sum of rings $A = A_1 \oplus A_2$ with $A_1 = Ae$ and $A_2 = A(1 - e)$.

Pedantry Notice that the inclusion $A_1 \hookrightarrow A_1 \oplus A_2$ is not a subring according to Convention 1.1, because the 1 of A_1 maps to $(1, 0)$, which is not the 1 of $A_1 \oplus A_2$. It is however, an ideal, even a principal ideal. Also, A_1 is a ring of fractions of $A_1 \oplus A_2$ (in the sense of Chapter 6) under the projection map $A_1 \oplus A_2 \twoheadrightarrow A_1$.

The point of these definitions is as follows. If a ring A is an integral domain, it obviously has no nilpotents or idempotents other than 0 and 1, and 0 is a prime ideal, and therefore the only minimal prime ideal. If A is not an integral domain, then it has zerodivisors $f, g \neq 0$ with $fg = 0$. There are two essentially different ways this can happen, exemplified by $k[X, Y]/(Y^2)$ and $k[X, Y]/(XY)$. In the first case $A = k[X, Y]/(Y^2)$ contains a nonzero element Y of square 0; that is, $fg = 0$ where f is contained in all the prime ideals P containing g (in the present instance, $f = g = Y$). In the second case the relation is $XY = 0$, where X and Y belong to different minimal prime ideals (X) and (Y). The ring $A = k[X, Y]/(XY)$ is a subring of the direct sum $k[X] \oplus k[Y]$, with X

and Y mapping to a non-zerodivisor in one factor and to zero in the other, so that their product is zero.

Proposition *Let A be a ring with zerodivisors. Then A has either nonzero nilpotent elements, or more than one minimal prime ideal.*

Proof I can assume that nilrad $A = 0$, that is, that A has no nilpotent elements. By Corollary 1.10, $0 = \bigcap P$, with the intersection taken over all prime ideals $P \in \operatorname{Spec} A$. However, if $P_1 \supset P_2$ are prime ideals then in taking this intersection, I can just omit P_1. On the other hand, it is easy to see that every prime ideal contains a minimal prime ideal (see Ex. 1.18); therefore $0 = \bigcap P$ with the intersection taken over all minimal primes. If there is only one minimal prime P then $P = 0$ and A is an integral domain. Q.E.D.

The proposition becomes more significant later in the book: by Corollary 5.13, a Noetherian ring has only finitely many minimal prime ideals P_i. If A is reduced, this means that it has an inclusion into a direct sum $A \hookrightarrow \bigoplus_{i=1}^{n} A/P_i$, just as in the above example $XY = 0$. See Ex. 1.13.

Geometrically, $k[X, Y]/(XY)$ is the ring of functions on the union of the two coordinate axes in k^2, given by $(XY = 0) \subset k^2$; its minimal prime ideals (X) and (Y) correspond to the two components, and the direct sum $k[X] \oplus k[Y]$ is the ring of functions on the disjoint union of the two axes.

The geometric picture of nilpotents is in the spirit of a nonrigorous introduction to calculus, where one may talk about a "number ε so small that $\varepsilon^2 = 0$", and use this to calculate the derivative $f'(x)$ of a polynomial by the formula $f(x + \varepsilon) = f(x) + \varepsilon f'(x)$. This is of course nonsense in the context of an element of the real field, but is perfectly sensible and useful in algebra. Thus the element $Y \in k[X, Y]/(Y^2)$ is pictured as the function on the X-axis with values in $k[\varepsilon]/(\varepsilon^2 = 0)$ which remembers the Y-derivative $(\partial f/\partial Y)(x, 0)$ of $f(X, Y) \in k[X, Y]$ at each point $(x, 0)$. Higher order nilpotents can be used to give an algebraic treatment of Taylor expansions of polynomials up to any order.

1.12 Radical of an ideal

Let I be an ideal of A. The *radical* of I is the set

$$\operatorname{rad} I = \{ f \in A \mid f^n \in I \text{ for some } n \};$$

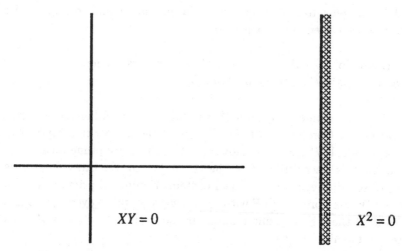

Figure 1.11. The plane curves defined by $XY = 0$ and $X^2 = 0$

(this is sometimes also written $\sqrt{I}$). Obviously, nilrad $A = \operatorname{rad} 0$. If $\varphi \colon A \to A/I$ is the quotient homomorphism, $\operatorname{rad} I$ consists exactly of the elements of A that map to nilpotents of A/I, that is

$$\operatorname{rad} I = \varphi^{-1}\big(\operatorname{nilrad}(A/I)\big);$$

in particular, of course, $\operatorname{rad} I$ is an ideal (you can of course check this directly). We say that I is a *radical* ideal if $I = \operatorname{rad} I$.

Corollary $\operatorname{rad} I$ *is the intersection of all prime ideals* P *of* A *containing* I, *that is*

$$\operatorname{rad} I = \bigcap_{\substack{P \in \operatorname{Spec} A \\ P \supset I}} P.$$

(Here if $I = A$ the empty intersection means the whole ring A.)

Proof Since the prime ideals of A containing I are exactly the ideals of the form $\varphi^{-1}(Q)$ with Q prime in A/I, the result comes on taking φ^{-1} of the two sides of $(*)$ in Corollary 1.10. Alternatively, it is easy to argue directly: if $f^n \notin I$ for any n then

$$S = \{1, f, f^2, \dots\}$$

is a multiplicative set disjoint from I, and therefore by Proposition 1.9, there exists a prime ideal P containing I and disjoint from S. In particular $f \notin P$.   Q.E.D.

Corollaries 1.10 and 1.12 are a rather easy analogue of the Nullstellensatz (see Theorem 5.6), a point I will discuss in detail in 5.14, (iv) below.

1.13 Local ring

I now introduce one of the key notions of the course. A ring is *local* if it has a unique maximal ideal m. By Corollary 1.9,

$$A \text{ is local} \iff A \text{ has only one maximal ideal}$$
$$\iff \text{all the nonunits of } A \text{ form an ideal.}$$

Another equivalent condition is that A has a maximal ideal m such that $1 + m \subset A^{\times}$; compare Ex. 1.10. One often writes $A \supset m$ or (A, m) or (A, m, k) for a local ring A, its maximal ideal m and the residue field $k = A/m$.

You're probably not yet aware of any local rings, other than fields. I introduce in Chapter 6 below the algebraic procedure of *localisation*, which constructs local rings out of any ring. For example, if P is a prime ideal of an integral domain A, and $K = \operatorname{Frac} A$ is the field of fractions of A, then

$$A_P = \left\{ f/g \in K \,\middle|\, \begin{matrix} f, g \in A \\ \text{with } g \notin P \end{matrix} \right\} \supset m_P = \left\{ f/g \in K \,\middle|\, \begin{matrix} f, g \in A \text{ with} \\ g \notin P, f \in P \end{matrix} \right\}$$

is a local subring of K. Indeed, allowing any $g \notin P$ to appear in denominators has the effect of sabotaging any ideal not contained in P. We will see many times in the rest of the course that localisation can be used to reduce all kinds of questions concerning arbitrary rings to local rings. See, for example, Proposition 8.7.

1.14 First examples of local rings

Here are two examples of localisation: suppose that in $\mathbb{Z}$, I am interested in divisibility by a particular prime number, say 5. Then for $n \in \mathbb{Z}$, obviously

$$n \text{ is divisible by 5 in } \mathbb{Z} \iff n \text{ is divisible by 5 in } \mathbb{Z}[1/2, 1/3, 1/7];$$

that is, allowing prime factors 2, 3 or 7 into the denominators has no effect on divisibility by 5. So I might as well go the whole hog, and allow any prime factor other than 5 in the denominator, setting

$$\mathbb{Z}_{(5)} = \{q \in \mathbb{Q} \mid q = a/b \text{ with } a, b \in \mathbb{Z}, 5 \nmid b\}.$$

Then it is easy to see that

$$q = a/b \text{ is a unit of } \mathbb{Z}_{(5)} \iff 5 \nmid a.$$

Therefore

$$\text{nonunits of } \mathbb{Z}_{(5)} = \{a/b \in \mathbb{Z}_{(5)} \mid 5 \mid a\} = 5\mathbb{Z}_{(5)};$$

this is an ideal, so that $\mathbb{Z}_{(5)}$ is a local ring with maximal ideal $5\mathbb{Z}_{(5)}$, and residue field $\mathbb{Z}_{(5)}/5\mathbb{Z}_{(5)} \cong \mathbb{Z}/(5) = \mathbb{F}_5$.

In the same way, if I am working in $k[X]$ and only interested in divisibility of polynomials by powers of X, then I can replace $k[X]$ by rings such as $k[X][1/(X+1), 1/(X^2+1)]$ in which some other polynomials are made invertible, or go the whole hog and set

$$k[X]_{(X)} = \{h \in k(X) \mid h = f/g \text{ with } f, g \in k[X], X \nmid g\}$$
$$= \{h \in k(X) \mid h = f/g \text{ with } f, g \in k[X], g(0) \neq 0\}.$$

Since f/g is invertible in $k[X]_{(X)}$ if $X \nmid f$, it follows that this is a local ring, with maximal ideal generated by X. Note that an element of $k[X]_{(X)}$ is a rational function whose denominator does not vanish at 0, so can be viewed as a function defined near 0.

1.15 Power series rings and local rings

Here are two examples of local rings appearing in other areas of human endeavour. First the formal power series ring $k[[X]]$, which is familiar from calculations with power series when you don't worry about convergence. Let k be a field, and define the formal power series ring in a variable X over k by

$$k[[X]] = \{\text{formal power series in } X \text{ with coefficients in } k\}$$
$$= \left\{\sum_{n=0}^{\infty} a_n X^n \mid \text{with } a_n \in k\right\}$$

(compare Ex. 0.12). If $f = a_0 + a_1 X + a_2 X^2 + \cdots$ is a formal power series then f has an inverse in $k[[X]]$ if and only if $a_0 \neq 0$. Because in this case $f = a_0(1 + Xg)$ with some $g \in k[[X]]$, and

$$f^{-1} = a_0{}^{-1}(1 - Xg + X^2g^2 - \cdots).$$

It is easy to check that this is a well-defined power series, since the coefficient of X^n comes only from the first $(n + 1)$ terms of the infinite sum. Therefore

$$f \in k[[X]] \text{ is a nonunit} \iff a_0 = 0 \iff f \in (X),$$

so that $k[[X]]$ is a local ring with maximal ideal (X).

Note that $k[X] \subset k[[X]]$, and, by what I just said, any polynomial $g \in k[X]$ with $g(0) \neq 0$ is invertible in $k[[X]]$, so that the local ring $k[X]_{(X)}$ described above can be viewed as a subring of the formal power series ring, $k[X]_{(X)} \subset k[[X]]$; this inclusion takes a rational function $h = f/g$ defined near 0 to its Taylor series at 0.

Now in the case $k = \mathbb{R}$ or $\mathbb{C}$, we can also discuss convergent power series. We know from analysis that a power series $f = \sum_{n=0}^{\infty} a_n X^n$ in a variable X has *radius of convergence* ρ if $|a_n| \leq \text{const} \cdot \rho^{-n}$ for all n. In other words, f has positive radius of convergence if and only if

$$\limsup \frac{\log |a_n|}{n} < \infty.$$

Then f is convergent on a small disc around 0, and can be viewed as an analytic function near 0.

Let $\mathbb{R}\{X\}$ or $\mathbb{C}\{X\}$ be the set of power series with radius of convergence > 0. It is easy to see from elementary properties of analytic functions that f^{-1} is an analytic function represented by a convergent power series if and only if $a_0 \neq 0$. Thus again $\mathbb{R}\{X\}$ or $\mathbb{C}\{X\}$ is a local ring with maximal ideal (X).

The local ring $\mathbb{C}\{X\}$ occurs in complex function theory as the ring of germs of analytic (= holomorphic) functions around 0; here *germ* means that every $f \in \mathbb{C}\{X\}$ is an analytic function on a neighbourhood U of 0 (the neighbourhood U depends on f). The field of fractions of $\mathbb{C}\{X\}$ is the field of Laurent power series, or germs of meromorphic functions around 0. For most algebraic purposes, $\mathbb{C}\{X\}$ is very similar to $\mathbb{C}[[X]]$.

The same ideas apply to continuous functions near the origin, compare 3.3.

Exercises to Chapter 1

1.1   Give an example of a ring A and ideals I, J such that $I \cup J$ is not an ideal; in your example, what is the smallest ideal containing I and J?

1.2   The *product* of two ideals I and J is defined as the set of all sums $\sum f_i g_i$ with $f_i \in I$, $g_i \in J$. Give an example in which $IJ \neq I \cap J$.

1.3   Let $A = k[X, Y]/(XY)$. Show that any element of A has a unique representation in the form

$$a + f(X)X + g(Y)Y \quad \text{with} \quad a \in k, f \in k[X], g \in k[Y].$$

How do you multiply two such elements?

Prove that A has exactly two minimal prime ideals. If possible, find ideals I, J, K to contradict each of the following statements:

(a) $IJ = I \cap J$;
(b) $(I + J)(I \cap J) = IJ$;
(c) $I \cap (J + K) = (I \cap J) + (I \cap K)$.

1.4   Two ideals I and J are *strongly coprime* if $I + J = A$. Check that this is the usual notion of coprime for $A = \mathbb{Z}$ or $k[X]$. Prove that if I and J are strongly coprime then

$$IJ = I \cap J \quad \text{and} \quad A/IJ \cong (A/I) \times (A/J).$$

Prove also that if I and J are strongly coprime then so are I^n and J^n for $n \geq 1$.

1.5   Let $\varphi \colon A \to B$ be a ring homomorphism. Prove that φ^{-1} takes prime ideals of B to prime ideals of A. In particular if $A \subset B$ and P is a prime ideal of B then $A \cap P$ is a prime ideal of A.

1.6   Prove or give a counterexample:

(a) the intersection of two prime ideals is prime;
(b) the ideal $P_1 + P_2$ generated by 2 prime ideals P_1, P_2 is again prime;
(c) if $\varphi \colon A \to B$ is a ring homomorphism then φ^{-1} takes maximal ideals of B to maximal ideals of A;
(d) the map φ^{-1} of Proposition 1.2 takes maximal ideals of A/I to maximal ideals of A.

1.7   Give an example of a ring A and a multiplicative set S such that $A \setminus S$ is not a prime ideal. Read again the definition of multiplicative set and prime ideal in 1.3, and check that all is right.

1.8   Prove that, as asserted in 0.2, the set nilrad A of nilpotent elements of A is an ideal; similarly, for any ideal I, the radical rad I (defined in 1.12) is again an ideal.

1.9   (a) If a is a unit and x is nilpotent, prove that $a + x$ is again a unit. [Hint: expand $(1 + x/a)^{-1}$ as a power series to guess the inverse.]

        (b) Let A be a ring, and $I \subset$ nilrad A an ideal; if $x \in A$ maps to an invertible element of A/I, prove that x is invertible in A.

1.10   Think through the equivalent conditions in the definition of local ring in 1.13. In particular, let A be a ring and I a maximal ideal such that $1 + x$ is a unit for every $x \in I$, and prove that A is a local ring. Does the implication still hold if I is not maximal?

1.11   Find the nilpotent and idempotent elements of $\mathbb{Z}/(n)$, where $n = 6$, 12, pq, pq^2 or $\prod p_i{}^{n_i}$ (where p, q, p_i are distinct prime numbers).

1.12   (a) If I and J are ideals and P a prime ideal, prove that

$$IJ \subset P \iff I \cap J \subset P \iff I \text{ or } J \subset P.$$

        (b) If $I, J_1, J_2 \subset A$ are ideals, prove that $I \subset J_1 \cup J_2$ implies either $I \subset J_1$ or $I \subset J_2$. If in addition P is a prime ideal and $I \subset J_1 \cup J_2 \cup P$, prove that $I \subset J_1$ or J_2 or P. [Hint: compare [A & M], Proposition 1.11.]

1.13   If A is a reduced ring and has finitely many minimal prime ideals P_i then $A \hookrightarrow \bigoplus_{i=1}^{n} A/P_i$; moreover, the image has nonzero intersection with each summand. Compare the discussion in 1.11.

1.14   Let $f = f(x_1, \ldots, x_n)$ be an analytic function (of real variables) in a neighbourhood of a point $P = (a_1, \ldots, a_n) \in \mathbb{R}^n$. By considering the Taylor series expansion of f about P, convince yourself that $f(a_1, \ldots, a_n) = 0$ if and only if f has an expression $f = \sum (x_i - a_i) \cdot g_i$ with g_i analytic functions near $a_1, \ldots, a_n$.

    Analytic functions near P form a ring. Deduce that the ideal of analytic functions vanishing at $P = (a_1, \ldots, a_n)$ is generated by $\{x_1 - a_1, \ldots, x_n - a_n\}$.

1.15   Let $a = (a_1, \ldots, a_n) \in k^n$, and consider the map

$$e_a \colon k[x_1, \ldots, x_n] \mapsto k$$

defined by $f \mapsto f(a_1, \ldots, a_n)$; in other words, e_a is the map

given by considering a polynomial $f(x)$ as a function and evalu-
ating it at $x = a$. Prove that $\ker e_a = (x_1 - a_1, \dots, x_n - a_n)$.
[Hint: do this in two steps: Step 1, the case $a = (0, \dots, 0)$.
Step 2, a coordinate change $y_i = x_i - a_i$.] Deduce that the
ideal $(x_1 - a_1, \dots, x_n - a_n)$ is a maximal ideal of $k[x_1, \dots, x_n]$.
Compare with the remainder theorem (see Ex. 0.2).

1.16   Now let $a = (a_1, \dots, a_n) \in K^n$, where $k \subset K$ is an algebraic field
        extension. Determine the image and kernel of the evaluation
        map $e_a \colon k[x_1, \dots, x_n] \mapsto K$ defined by $f \mapsto f(a_1, \dots, a_n)$. Prove
        that $(x_1 - a_1, \dots, x_n - a_n) \cap k[x_1, \dots, x_n]$ is a maximal ideal of
        $k[x_1, \dots, x_n]$ (the intersection takes place in $K[x_1, \dots, x_n]$).

1.17   Describe $\operatorname{Spec} \mathbb{R}[X]$ in terms of $\mathbb{C}$. [Hint: list the irreducible
        polynomials $f \in \mathbb{R}[X]$, and say how they factorise in $\mathbb{C}[X]$.]

1.18   Use Zorn's lemma to prove that any prime ideal P contains a
        minimal prime ideal.

1.19   Use Zorn's lemma to prove the following version of the axiom
        of choice: a surjective map $f \colon X \to Y$ between sets has a right
        inverse $g \colon Y \to X$ such that $f \circ g = \mathrm{id}_Y$; in other words, it
        is possible to choose one inverse image for each $y \in Y$ all at
        the same time. [Hint: let Σ be the set of pairs (U, γ), where
        $U \subset Y$ and $\gamma \colon U \to X$ is a map such that $f(\gamma(y)) = y$ for all
        $y \in U$; introduce a suitable partial order $\leq$ on Σ (the "obvious"
        one), prove that it satisfies the assumptions of Zorn's lemma,
        and show that a maximal element of Σ does what I said.]

1.20   Sūn WùKòng, The Monkey King in Wu Cheng-En's famous
        novel "Journey to the West", has "Aware of Vacuity" as his
        given name (Arthur Waley's translation). Show that $\operatorname{Spec} A = \emptyset$
        if and only if $A = \{0\}$ (compare 1.3). In this case, describe how
        to view A as the ring of functions on $\operatorname{Spec} A$ (there's really
        nothing in it!).

2
Modules

2.1 Definition of a module

A is a ring; an A-*module* is an Abelian group M with a multiplication map

$$A \times M \to M, \quad \text{written } (f, m) \mapsto fm$$

satisfying

(i) $f(m \pm n) = fm \pm fn$;
(ii) $(f + g)m = fm + gm$;
(iii) $(fg)m = f(gm)$;
(iv) $1_A m = m$

for all f, $g \in A$ and m, $n \in M$. A subset $N \subset M$ is a *submodule* if $fm + gn \in N$ for all f, $g \in A$ and m, $n \in N$. A *homomorphism* is a map $t\colon M \to N$ of A-modules that is A-linear, in the obvious sense that $t(fm + gn) = ft(m) + gt(n)$ for all $f, g \in A$ and $m, n \in M$.

Note that a module over a field k is just a k-vector space: the definition is word-for-word the same. Thus the theory of modules is just linear algebra in which scalar multiplication is by elements of A, and linear combinations $\sum a_i m_i$ take place with $m_i \in M$ and coefficients $a_i \in A$. This allows us to do linear algebra over any ring A (which doesn't even have to be commutative in general).

2.2 Harmless formalism

Introduce the temporary notation $\mu_f \colon M \to M$ "multiplication by f" for the map $m \mapsto fm$; then you can see at once that (i) holds if and only if μ_f is a homomorphism of Abelian groups and (ii–iv) hold if and only if the map $f \mapsto \mu_f$ is a ring homomorphism $A \to \operatorname{End} M$ from A to the noncommutative ring of endomorphisms of M (that is, homomorphisms of the Abelian group M to itself). Notice that μ_f is A-linear (that is, $f(am + bn) = a(fm) + b(fn)$), and that this fact depends on the assumption that A is commutative.

As far as M is concerned, $\mu(A) \subset \operatorname{End} M$ is the only part of A that matters, and I sometimes write $A' = \mu(A)$. If μ is injective, then $A \cong A'$, and linear algebra in M reflects all the properties of A; in traditional algebraic terminology, M is a *faithful* A-module.

If M is an A-module and $\varphi \colon M \to M$ an A-linear endomorphism of M, I write $A'[\varphi] \subset \operatorname{End} M$ for the subring generated by A and the action of φ, that is, the ring generated by μ_f for $f \in A$ and φ; this is a commutative ring, since φ is A-linear. This formal construction makes M a module over $A'[\varphi]$, so that the map φ becomes multiplication by a ring element. Notice that multiplication in the ring $A'[\varphi] \subset \operatorname{End} M$ coincides with composition of maps $M \to M$. This notation will be applied in 2.6–2.7 below.

Examples A ring A is a module over itself, and a submodule $I \subset A$ is the same thing as an ideal. An Abelian group is a $\mathbb{Z}$-module in an obvious way; in fact the rules for manipulating modules in general are in many respects based on those for Abelian groups. If A is a subring $A \subset B$ of a ring B then multiplication in B makes B into an A-module; similarly, a B-module M can be viewed as an A-module by just restricting scalar multiplication to elements of A.

The most important modules in commutative algebra are those closely related to the ring itself, for example, ideals of A or extension rings $B \supset A$, or sums and quotients of these. However, there is no advantage in working with a more restricted class of modules.

2.3 The homomorphism and isomorphism theorems

The proof of the following proposition is exactly the same as for vector spaces, and I do not spend time on it.

Proposition

(1) *The kernel $\ker \varphi \subset M$ and image $\operatorname{im} \varphi \subset N$ of an A-module homomorphism $\varphi \colon M \to N$ are submodules.*

(2) *If $N \subset M$ is a given submodule, there exists a quotient module M/N and a surjective quotient map $\varphi \colon M \to M/N$ such that*

$\ker \varphi = N$. *The elements of M/N can be constructed as equivalence classes of elements $m \in M$ modulo N, or as cosets $m + N$ of N in M.*

(3) *In (1), $M/\ker \varphi \cong \operatorname{im} \varphi$.*

Isomorphism theorems

(i) *If $L \subset M \subset N$ are submodules then*

$$N/M = (N/L) \big/ (M/L).$$

(ii) *If N is a module, and $L, M \subset N$ are submodules then*

$$(M + L)/L = M/(M \cap L).$$

In either case, equality stands for a canonical isomorphism.

Note that (II) can be interpreted as saying that if $L \not\subset M$ then there are two possible ways of making sense of the expression M/L, by increasing M until it contains L, or by decreasing L until it is contained in M, and these both give the same result.

Proof Both of these follow from the Proposition, (3) via an argument of the form: Step 1, construct a map; and Step 2, calculate its kernel.

(I) First of all, I claim that there is a well-defined homomorphism $\varphi \colon N/L \to N/M$. Indeed, take some element $a \in N/L$; since $N \to N/L$ is surjective, there is an $n \in N$ mapping to a, and I set

$$\varphi(a) = \text{image of } n \text{ in } N/M;$$

this is well defined because $L \subset M$. It is easy to check that φ is A-linear. Second, I claim that $\ker \varphi = M/L$, since clearly

$$\varphi(a) = 0 \iff n \in M \iff a \in M/L.$$

Therefore I have $\varphi \colon N/L \to N/M$ with $\ker \varphi = M/L$, so that the result follows from (3) of the Proposition.

The proof of (II) is similar: the composite $\varphi \colon M \to (M + L)/L$ of the inclusion $M \subset M + L$ and the projection $M + L \to (M + L)/L$ is A-linear and satisfies $\ker \varphi = L \cap M$. Q.E.D.

2.4 Generators of a module

Given $m_1, \ldots, m_r \in M$, I can form the submodule of linear combinations

$$(m_1, \ldots, m_r) = \sum A m_i = \left\{ \sum f_i m_i \in M \,\middle|\, f_i \in A \right\} \subset M,$$

and argue about whether this is the whole of M, and what relations there are between the m_i; in the case of linear algebra over a field, this is the theory of linear independence and bases of vector spaces. I formulate the definition with some care to include infinite sets of generators $(m_\lambda)_{\lambda \in \Lambda}$, but for many purposes this is not needed, and you would do well to concentrate on the case $\Lambda = \{1, \ldots, n\}$ in the first instance. A typical example where infinite sets of generators might be useful is a polynomial ring such as $A[X, Y]$, which is a free A-module with the countable basis $X^i Y^j$.

If $(M_\lambda)_{\lambda \in \Lambda}$ is a family of A-modules, their *direct sum* is defined by

$$\sum_{\lambda \in \Lambda} M_\lambda = \left\{ (m_\lambda)_{\lambda \in \Lambda} \,\middle|\, \text{only finitely many } m_\lambda \neq 0 \right\}.$$

The indexing set of the sum is infinite, but each element of the sum involves only finitely many indexes. In particular, I can take the sum of A itself card Λ times:

$$A^{\operatorname{card} \Lambda} := \sum_{\lambda \in \Lambda} A = \left\{ (a_\lambda)_{\lambda \in \Lambda} \,\middle|\, \text{only finitely many } a_\lambda \neq 0 \right\}.$$

in other words, this is the set of (finite) sequences indexed by $\lambda \in \Lambda$. This is by definition a *free module*; it has a *basis* $e_\lambda = (0, \ldots, 1, \ldots)$, with 1 in the λth place and zero elsewhere. In particular, if $(m_\lambda)_{\lambda \in \Lambda}$ is any set of elements of a fixed module M indexed by Λ, I can define a homomorphism from this direct sum to M,

$$\varphi \colon A^{\operatorname{card} \Lambda} \to M \qquad \text{by } (f_\lambda)_{\lambda \in \Lambda} \mapsto \sum f_\lambda m_\lambda;$$

the sum is finite since every element $(f_\lambda) \in A^{\operatorname{card} \Lambda}$ has only finitely many $f_\lambda \neq 0$. The image is obviously just the set of all finite linear combinations $\sum f_\lambda m_\lambda$ of the given elements m_λ.

Definition $(m_\lambda)_{\lambda \in \Lambda}$ is a *family of generators* of M if φ is surjective, that is, if any $m \in M$ can be written $m = \sum f_\lambda m_\lambda$; it is a *basis* if φ is an isomorphism. M is *free* if it has a basis. A module M is *finite* over A if it has a finite set of generators. Note that the term finite is used throughout to mean finitely generated as an A-module.

2.5 Examples

Formally speaking, the definition just given is exactly like that of a basis of a vector space over a field. However, every vector space (say, finite dimensional) has a basis: every irredundant set of generators is a basis, and so is every maximal set of linearly independent elements. None of this is true for modules over a ring A, as illustrated by the following examples, which can be summed up in the slogan "a free module is a lucky accident". However, free modules are also important because a general module can be studied in terms of them using a presentation by generators and relations (see, for example, Ex. 3.4).

(1) If A is an integral domain and $0 \neq f \in A$ then $A[1/f]$ is usually not finitely generated as an A-module: $A[1/f] = A + Af^{-1} + Af^{-2} + \cdots$, so $\{1, f^{-1}, f^{-2}, \dots\}$ is a set of generators. See Ex. 2.1.

(2) The module A/I is generated by 1 element (the image of 1_A), but is not free if $I \neq 0$; a module of the form A/I is sometimes called a *cyclic* module.

(3) In $A = k[X, Y]$, the maximal ideal $M = (X, Y)$ is generated by two elements, but is not a free module: the map $\varphi \colon A^2 \to M$ is given by $(f, g) \mapsto fX + gY$, and has kernel $\{(hY, -hX) \mid h \in A\}$. See Ex. 2.2

(4) Even when M is free, an irredundant family of generators is not necessarily a basis; for example, $A = k[X]$, $M = A$ and $(X, 1-X)$.

2.6 The Cayley–Hamilton theorem

If N is an $n \times n$ matrix over an algebraically closed field k, considered as a endomorphism $N \colon k^n \to k^n$, then if N is "fairly general", it can be diagonalised; that is, N can be written as the diagonal matrix $N = \operatorname{diag}(\lambda_1, \dots, \lambda_n)$ by choosing a basis of k^n made up of eigenvectors e_i. Now $(N - \lambda_i \cdot I_n)e_i = 0$, where I_n is the identity $n \times n$ matrix, and it follows that the product of the n factors $(N - \lambda_i \cdot I_n)$ kills every basis element e_i, so that

$$(N - \lambda_1 \cdot I_n) \cdots (N - \lambda_n \cdot I_n) = 0 \tag{1}$$

(the factors commute, of course). On the other hand, $p_N(x) = \prod(x - \lambda_i)$ is the characteristic polynomial of N, that is, $p_N = \det(x \cdot I_n - N)$, and hence (1) says that $p_N(N) = 0$. The Cayley–Hamilton theorem says that the same thing holds for any $n \times n$ matrix with entries in any (commutative) ring A.

Theorem *Let A be a commutative ring, $N = (a_{ij})$ an $n \times n$ matrix with entries in A, and write $p_N(x) = \det(x \cdot I_n - N)$. This is a polynomial of degree n in x with coefficients in A and leading coefficient 1.*

Then $p_N(N) = 0$ (as an $n \times n$ matrix).

Note that this is an array of n^2 identities in the entries a_{ij} of N. You'll get some feel for the statement by writing it out explicitly for $n = 2, 3, \ldots$; see Ex. 2.4.

Proof Write $\varphi \colon A^n \to A^n$ for the homomorphism of the free module A^n corresponding to N, that is, the map taking

$$e_k = (0, \ldots, 1, \ldots, 0) \mapsto (0, \ldots, 1, \ldots, 0)N = (a_{k1}, \ldots, a_{kn}), \quad (2)$$

the kth row of N. Let $A'[\varphi]$ be the commutative subring of $\operatorname{End} A^n$ discussed in 2.2, that is, the subring of $n \times n$ matrixes generated by the scalar multiples of I_n and N.

Viewing A^n as an $A'[\varphi]$-module in the natural way, I get the equations

$$\varphi e_k = \sum_{i=1}^{n} a_{ki} e_i \qquad \text{for } k = 1, \ldots, n, \quad (3)$$

which I can rewrite

$$\sum_{i=1}^{n} (\varphi \delta_{ki} - a_{ki}) e_i = 0 \quad (\delta_{ki} \text{ is the Kronecker symbol}). \quad (4)$$

Now let $\Delta = (\varphi \delta_{ki} - a_{ki})$, which is an $n \times n$ matrix with coefficients in $A'[\varphi]$.

I claim that $\det \Delta = 0$ in $A'[\varphi]$. Since $p_N(x) = \det(x \cdot I_n - N) \in A[x]$ and $A[x] \to A'[\varphi]$ is a homomorphism (so commutes with the algebraic operations involved in taking determinant), it follows that $p_N(x)$ maps to $p_N(N) = \det \Delta$, so that the claim implies the theorem.

To prove the claim, let $\operatorname{adj} \Delta = (b_{kj})$ be the adjoint matrix of Δ. This is the matrix whose entry b_{kj} is the (j, k)th factor of Δ (that is, $(-1)^{j+k}$ times the $(n-1) \times (n-1)$ determinant obtained by deleting the jth row and kth column of Δ), and

$$(\operatorname{adj} \Delta) \cdot \Delta = \Delta \cdot (\operatorname{adj} \Delta) = \det \Delta \cdot I_n \quad (5)$$

is just the familiar rule for evaluating a determinant along the rows or columns. Now $\det \Delta \in A'[\varphi] \subset \operatorname{End} A^n$, so that it is enough to prove that $(\det \Delta) \cdot e_j = 0$ for $j = 1, \ldots, n$. For this, just multiply (4) through by b_{kj} and sum over k. This gives

$$\sum_{k=1}^{n} b_{kj} \cdot \sum_{i=1}^{n} (\varphi \delta_{ki} - a_{ki}) e_i = 0. \tag{6}$$

By (5), the coefficient before e_i is just $(\det \Delta)\delta_{ji}$; in other words, I have proved that $(\det \Delta)e_j = 0$ for each j, as required. Q.E.D.

2.7 The determinant trick

The following result is proved in exactly the same way as the Cayley–Hamilton theorem, and has many applications later in the course (see Proposition 4.2, (iii) $\implies$ (i) and the Remark following Proposition 4.2).

Theorem *Let M be a finite A-module, generated by n elements, and $\varphi \colon M \to M$ a homomorphism; suppose that I is an ideal of A such that $\varphi(M) \subset IM$. Then φ satisfies a relation of the form*

$$\varphi^n + a_1\varphi^{n-1} + \cdots + a_{n-1}\varphi + a_n = 0, \tag{7}$$

where $a_i \in I^i$ for $i = 1, 2, \ldots, n$. Here (7) is an equation holding between endomorphisms of M, that is, a relation in the ring $A'[\varphi] \subset \operatorname{End} M$ introduced in 2.2.

Proof Let $m_1, \ldots, m_n$ be a set of generators of M. Since $\varphi(m_i) \in IM$, I can write

$$\varphi(m_i) = \sum a_{ij}m_j \quad \text{with } a_{ij} \in I. \tag{8}$$

As before, in the ring $A'[\varphi]$, (8) can be rewritten

$$\sum (\delta_{ij}\varphi - a_{ij})m_j = 0, \tag{9}$$

with $\delta_{ij}\varphi - a_{ij} \in A'[\varphi]$. Write $\Delta = (\delta_{ij}\varphi - a_{ij})$. Then multiplying (9) by b_{ki} and summing over i, I deduce that $(\det \Delta)m_k = 0$. Since this holds for each k, it follows that $\det \Delta = 0 \in A'[\varphi]$. Expanding out the determinant gives the relation in the statement. Q.E.D.

2.8 Corollaries – Nakayama's lemma

Corollary 1 *If M is a finite module and $M = IM$ then there exists an element $x \in A$ such that $x \equiv 1 \bmod I$ and $xM = 0$.*

Proof Apply the determinant trick 2.7 to the identity map $\mathrm{id}_M \in A'[\varphi]$. Since $(1_M)^i = 1_M$, the relation takes the form $(1 + b)\mathrm{id}_M = 0$, where $b = a_1 + \cdots + a_n \in I$.   Q.E.D.

Corollary 2 (Nakayama's lemma) *Let (A, m) be a local ring (defined in 1.13), and M a finite A-module; then $M = mM$ implies that $M = 0$. More generally, the same conclusion holds if $M = IM$ with I an ideal such that $(1 + I) \subset A^\times$.*

This is a simple but important result that allows us to make many deductions about properties of M as an A-module from properties of M/mM as a vector space over the residue field $k = A/m$, so reducing the study of finite modules over a local ring to finite dimensional vector spaces. It is not hard to prove directly, without using Theorem 2.7; for this, see Ex. 2.5.

Proof Apply the previous corollary. Then since $x \equiv 1 \bmod I$ it follows that x is a unit of A, so that $xM = 0$ implies $M = x^{-1} \cdot xM = 0$.   Q.E.D.

Corollary 3 *Let (A, m) be a local ring, M an A-module, and $N \subset M$ a submodule; suppose that M/N is finite over A, and that $M = N + mM$; then $N = M$.*

In particular, if M is finite over A, and if $s_1, \ldots, s_n \in M$ are elements whose images span the k-vector space $\overline{M} = M/mM$, then $s_1, \ldots, s_n$ generate M.

Proof Since $m(M/N) = M/N$, Nakayama's lemma (Corollary 2.8.2) gives $M/N = 0$.   Q.E.D.

Note that the finiteness condition is essential; for example, let $A = \mathbb{Z}_{(5)} \subset \mathbb{Q}$ be the example of local ring given in 1.14, with maximal ideal $5\mathbb{Z}_{(5)}$, and consider $\mathbb{Q}$ as A-module. Then obviously $\mathbb{Q} = 5\mathbb{Q}$, but $\mathbb{Q} \neq 0$; the point is that $\mathbb{Q} = A + A\frac{1}{5} + A\frac{1}{5^2} + \cdots$ is not finite over A.

2.9 Exact sequences

Suppose that L, M and N are A-modules, and

$$L \xrightarrow{\alpha} M \xrightarrow{\beta} N \qquad (9)$$

is a sequence of homomorphisms. (9) is *exact* at M if $\mathrm{im}\,\alpha = \ker \beta$; this means that the composite $\beta \circ \alpha = 0$, and that α maps surjectively to ker β. A longer sequence $\cdots \to M_1 \to M_2 \to M_3 \to \cdots$ is exact if it is exact at each term.

Special cases:

$$0 \to L \xrightarrow{\alpha} M \text{ is exact} \iff \alpha \text{ is injective;}$$

$$M \xrightarrow{\beta} N \to 0 \text{ is exact} \iff \beta \text{ is surjective;}$$

$$0 \to L \xrightarrow{\alpha} M \xrightarrow{\beta} N \text{ is exact} \iff \alpha\colon L \xrightarrow{\sim} \ker\beta;$$

$$L \xrightarrow{\alpha} M \xrightarrow{\beta} N \to 0 \text{ is exact} \iff \beta \text{ induces } M/\alpha(L) \xrightarrow{\sim} N.$$

By analogy with the kernel of a map, one defines the *cokernel* of a homomorphism $\alpha\colon L \to M$ to be the module $\operatorname{coker}\alpha = M/\alpha(L)$; then the conclusion in the final case can be written $\beta\colon \operatorname{coker}\alpha \xrightarrow{\sim} N$.

Most importantly,

$$0 \to L \xrightarrow{\alpha} M \xrightarrow{\beta} N \to 0 \text{ is exact} \iff L \subset M \text{ and } N = M/L,$$

under the natural identifications induced by α and β. A sequence with this property is called a *short exact sequence* (s.e.s.).

It would be possible to replace exact sequences by more direct arguments wherever they occur in this course, but I use them as a convenient language, which you probably need to become familiar with in any case. Treating exact sequences in a systematic way is the main subject of *homological algebra*, which is a major item of technology in 20th century math.

2.10 Split exact sequences

Proposition–Definition *Let*

$$0 \to L \xrightarrow{\alpha} M \xrightarrow{\beta} N \to 0$$

be a s.e.s. of A-modules. Then 3 equivalent conditions:

(i) *there exists an isomorphism $M \cong L \oplus N$ under which α is given by $m \mapsto (m, 0)$ and β by $(m, n) \mapsto n$;*

(ii) *there exists a section of β, that is, a map $s\colon N \to M$ such that $\beta \circ s = \operatorname{id}_N$;*

(iii) *there exists a retraction of α, that is, a map $r\colon M \to L$ such that $r \circ \alpha = \operatorname{id}_L$.*

If this happens the sequence is a split exact sequence.

Proof (i) $\implies$ (ii) or (iii) is very easy.

(ii) $\implies$ (i): The given section s is clearly injective because it has a left inverse; I claim that $M = \alpha(L) \oplus s(N)$. To see this, any $m \in M$ is of the form

$$m = \big(m - s(\beta(m))\big) + s(\beta(m)),$$

where the second term is obviously in $s(N)$; since $\beta \circ s = \mathrm{id}_N$, the first term is clearly in $\ker \beta$, and by exactness this is $\alpha(L)$. Furthermore, $\alpha(L) \cap s(N) = 0$, since if $n \in N$ is such that $s(n) \in \alpha(L) = \ker \beta$ then $n = \beta(s(n)) = 0$.

(iii) $\implies$ (i) is very similar, and I leave it to you as an exercise (Ex. 2.13). Q.E.D.

For finite dimensional vector spaces over a field, every subspace has a complement, so every s.e.s. is split. Whether an exact sequence is split or not depends on what ring it is considered over. For example, $0 \to (X) \to k[X] \to k \to 0$ is split over k but not over $k[X]$.

Exercises to Chapter 2

2.1　Let A be an integral domain with field of fractions K, and suppose that $f \in A$ is nonzero and not a unit. Prove that $A[1/f]$ is not a finite A-module. [Hint: if it has a finite set of generators then prove that $1, f^{-1}, f^{-2}, \ldots, f^{-k}$ is a set of generators for some $k > 0$, so that $f^{-(k+1)}$ can be expressed as a linear combination of these. Use this to prove that f is a unit.]

2.2　Let A be a UFD and $x, y \in A$ two elements having no common factor; write $I = (x, y) \subset A$. Prove that $\varphi \colon A^2 \to I$ defined by $(a, b) \mapsto ax + by$ is surjective and has kernel the submodule generated by the element $(-y, x)$. In other words, there is an exact sequence

$$0 \to A \xrightarrow{(-y,x)} A^2 \xrightarrow{\binom{x}{y}} I \to 0.$$

This is called the *Koszul complex* of the pair (x, y).

It can obviously happen that $I = (x, y) \neq (1)$ (for example, take $X, Y \in A = k[X, Y]$). Then I needs two generators, and is not a free module.

2.3　Let $A = k[X, Y, Z]$ be the polynomial ring. The maximal ideal $m = (X, Y, Z)$ is generated by the 3 elements X, Y, Z, so that there is a surjective homomorphism $\alpha \colon A^3 \to m \subset A$ defined by $(f, g, h) \mapsto fX + gY + hZ$. Prove that the kernel of α can be generated by 3 elements $u, v, w \in A^3$. Find the kernel of the surjective map $\beta \colon A^3 \to \ker \alpha$, and deduce that there is an exact sequence

$$0 \to A \to A^3 \xrightarrow{\beta} A^3 \xrightarrow{\alpha} m \to 0.$$

[Hint: as in Ex. 2.2, the only relations are stupid skewsymmetries.]

2.4　Verify the Cayley–Hamilton theorem (see 2.6) for $n = 2$ and 3; if you have access to a computer algebra system, do the same for $n = 4$ and 5.

2.5　Prove Nakayama's lemma from first principles, by induction on the minimum number of generators of M.

2.6　(A, m, k) is a local ring, M a finite A-module and $s_1, \ldots, s_n \in M$; write $\bar{s}_1, \ldots, \bar{s}_n \in \overline{M} = M/mM$ for their images under the quotient map. Prove that $\bar{s}_1, \ldots, \bar{s}_n$ form a basis of $\overline{M}$ if and only if $s_1, \ldots, s_n$ form a minimal generating set of M.

2.7　A is a local ring; prove that A^n and A^m are isomorphic as A-modules if and only if $n = m$. Now apply the existence of maximal ideals 1.8 to prove the same for any (commutative) ring A.

2.8　If A is a ring, and I a finitely generated ideal satisfying $I = I^2$, prove that I is generated by a single idempotent element. [Hint: you have to take the statement of Corollary 2.8.1 very literally.]

2.9　$0 \to L \to M \to N \to 0$ is a s.e.s. of A-modules; prove that if N and L are finite over A, then so is M.

2.10　$M_1, M_2 \subset N$ are submodules of a given A-module N; prove that if $M_1 + M_2$ and $M_1 \cap M_2$ are finite over A then so are M_1 and M_2. [Hint: apply Ex. 2.9 to a suitable s.e.s.]

2.11　Let $\cdots \to M_{r-1} \xrightarrow{\alpha_r} M_r \xrightarrow{\alpha_{r+1}} M_{r+1} \to \cdots$ be an exact sequence of modules; by setting $L_r = \operatorname{im} \alpha_r \subset M_r$, show how to break up any long exact sequence into a number of s.e.s.'s $0 \to L_r \to M_r \to L_{r+1} \to 0$.

2.12　Prove that if $0 \to V_1 \to V_2 \to \cdots \to V_n \to 0$ is an exact sequence of vector spaces over a field k then $\sum (-1)^i \dim V_i = 0$; this allows you to calculate the dimension of one of them if you know all the others. [Hint: you already know the case $n = 3$; now use Ex. 2.11 and induction.]

2.13　Prove (iii) implies (i) in Proposition 2.10.

2.14   Criticise the following misstatement of Proposition 2.10: given a s.e.s. $0 \to L \xrightarrow{\alpha} M \xrightarrow{\beta} N \to 0$,

$$M \cong L \oplus N \iff \exists \text{ section } s \colon N \to M \text{ of } \beta.$$

2.15   Prove the following lemma about exact sequences: let

$$
\begin{array}{ccccccccc}
0 \to & M_1 & \to & M_2 & \to & M_3 & \to 0 \\
& \downarrow \alpha_1 & & \downarrow \alpha_2 & & \downarrow \alpha_3 \\
0 \to & N_1 & \to & N_2 & \to & N_3 & \to 0
\end{array}
$$

be a commutative diagram of A-modules with exact rows; then α_1 and α_3 isomorphisms implies α_2 an isomorphism.

2.16   Give your own proof of the Snake lemma: if

$$
\begin{array}{ccccccc}
& M_1 & \to & M_2 & \to & M_3 & \to 0 \\
& \downarrow \alpha_1 & & \downarrow \alpha_2 & & \downarrow \alpha_3 \\
0 \to & N_1 & \to & N_2 & \to & N_3
\end{array}
$$

is a commutative diagram of A-modules with exact rows then there is a "snake" homomorphism $\delta \colon \ker \alpha_3 \to \operatorname{coker} \alpha_1$ (the cokernel is defined in 2.9) that fits together with the induced homomorphisms $\ker \alpha_1 \to \ker \alpha_2 \to \ker \alpha_3$ and $\operatorname{coker} \alpha_1 \to \operatorname{coker} \alpha_2 \to \operatorname{coker} \alpha_3$ to make a six term exact sequence

$$\ker \alpha_1 \to \ker \alpha_2 \to \ker \alpha_3$$
$$\xrightarrow{\delta} \operatorname{coker} \alpha_1 \to \operatorname{coker} \alpha_2 \to \operatorname{coker} \alpha_3.$$

[Hint: see for example [A & M], Proposition 2.10, or any book on algebraic topology or homological algebra.]

2.17   Give an alternative statement and proof of the first isomorphism theorem, Theorem 2.3, (I) in terms of exact sequences and the Snake lemma:

$$
\begin{array}{ccccccc}
& L & = & L & \to & 0 \\
& \downarrow & & \downarrow & & \downarrow \\
0 \to & M & \to & N & \to & N/M & \to 0 \\
& \downarrow & & \downarrow & & \downarrow \\
0 \to & M/L & \to & N/L & \to & \operatorname{coker} & \to 0 \\
& \downarrow & & \downarrow & & \downarrow \\
& 0 & & 0 & & 0
\end{array}
$$

3
Noetherian rings

In a paper of 1890, Hilbert proved that any ideal of a polynomial ring $k[X_1, \ldots, X_n]$ over a field is finitely generated, and used this result subsequently in 1893 in his proof that rings of invariants are finitely generated. His abstract methods of proving the existence of finite sets of generators without explicitly constructing them were denounced at the time as "theology, not mathematics". Work of Emmy Noether in the 1920s introduced the ascending chain condition, and used it systematically as a very convenient finiteness condition on a ring (see for example Theorem 7.11 below). Practically all the rings ever used in applications of commutative algebra are Noetherian. The heart's desire of any true algebraist, faced with a result on rings and modules that uses concrete assumptions, is to generalise the argument to work for all Noetherian rings.

3.1 The ascending chain condition

A partially ordered set Σ has the *ascending chain condition* (a.c.c.) if every chain

$$s_1 \leq s_2 \leq \cdots \leq s_k \leq \cdots$$

eventually breaks off, that is, $s_k = s_{k+1} = \cdots$ for some k. This is a finiteness condition in logic that allows arguments by induction, even when the partially ordered set Σ is infinite. It is easy to see that a partially ordered set Σ has the a.c.c. if and only if every nonempty subset $S \subset \Sigma$ has a maximal element: if $\emptyset \neq S \subset \Sigma$ does not have a maximal element, then choose $s_1 \in S$, and for each s_k, an element s_{k+1} with $s_k < s_{k+1}$, thus contradicting the a.c.c. For details see Ex. 3.10.

Vector subspaces of a finite dimensional vector space satisfy the a.c.c.:

the length of a strictly increasing chain is bounded by the dimension. The negative integers satisfy the a.c.c.: after choosing $-10,000,000$, say, the length of any strictly increasing chain is bounded. It is perhaps superfluous to remark that the a.c.c. does not say that the set Σ is finite, nor that the length of a chain is bounded, nor even that after making some finite number of choices $s_1 \leq \cdots \leq s_k$ the length of chains after s_k is bounded.

3.2 Noetherian rings

Proposition–Definition *Let A be a ring; 3 equivalent conditions:*

(i) *The set Σ of ideals of A has the a.c.c.; in other words, every increasing chain of ideals*

$$I_1 \subset I_2 \subset \cdots \subset I_k \subset \cdots$$

eventually stops, that is $I_k = I_{k+1} = \cdots$ for some k.

(ii) *Every nonempty set S of ideals has a maximal element.*

(iii) *Every ideal $I \subset A$ is finitely generated.*

If these conditions hold then A is Noetherian.

Proof Here (i) $\Longleftrightarrow$ (ii) is the purely logical statement about partially ordered sets already discussed, whereas (i) or (ii) $\Longleftrightarrow$ (iii) is directly concerned with rings and ideals.

(i) $\Longrightarrow$ (iii) Pick $f_1 \in I$, then if possible $f_2 \in I \setminus (f_1)$, and so on. At each step, if $I \neq (f_1, \ldots, f_k)$, pick $f_{k+1} \in I \setminus (f_1, \ldots, f_k)$. Then by the a.c.c. (i), the chain of ideals

$$(f_1) \subset (f_1, f_2) \subset \cdots \subset (f_1, \ldots, f_k) \subset \cdots$$

must break off at some stage, and this can only happen if $(f_1, \ldots, f_k) = I$. This proof involves an implicit appeal to the axiom of choice. It is perhaps cleaner to do (i) $\Longrightarrow$ (ii) purely in set theory, then argue as follows.

(ii) $\Longrightarrow$ (iii) Let I be an ideal of A, and consider the set S of finitely generated ideals contained in I. Then $0 \in S$, so that S has a maximal element J by (ii). But then $J = I$, since any element $f \in J \setminus I$ would give rise to a strictly bigger finitely generated ideal $J \subsetneq (J, f) \subset I$.

(iii) $\Longrightarrow$ (i) Let $I_1 \subset I_2 \subset \cdots \subset I_k \subset \cdots$ be an increasing chain. Then you check at once that $J = \bigcup I_k$ is again an ideal. If J is finitely generated then $J = (f_1, \ldots, f_n)$ and each $f_i \in I_{k_i}$, so that setting $k = \max k_i$ gives $J = I_k$, and the buck stops here.   Q.E.D.

Remarks

(i) Most rings of interest are Noetherian; this is a very convenient condition to work with. At first sight, more concrete conditions (such as A finitely generated over k or over $\mathbb{Z}$) might seem more attractive, but as a rule, the Noetherian condition is both more general and more practical to work with.

(ii) The descending chain condition (d.c.c.) on a partially ordered set is defined in a similar way; for example, the set of natural numbers $\mathbb{N}$ with the usual order satisfies the d.c.c. A ring whose ideals satisfy the d.c.c. is called an *Artinian ring*. This is also a very important notion, but is more special: the d.c.c. for rings turns out to be very much stronger than the a.c.c. (and implies it). See Exs. 3.8–9, as well as [A & M], Chapter 8 or [M], Chapter 1, Theorem 3.2.

3.3 Examples

I will give lots of examples of Noetherian rings later. Here are three examples of non-Noetherian rings:

(1) The polynomial ring $k[X_1, \ldots, X_n, \ldots]$ in an infinite number of indeterminates is obviously non-Noetherian.

(2) Consider the ring A_1 of polynomials in x, y of the form $f(x, y) = a + xg(x, y)$ with a a constant and $g \in k[x, y]$; that is, f involves no pure power y^j of y with $j > 0$. In other words,

$$A_1 = \left\{ f(x, y) = \sum a_{ij} x^i y^j \mid i, j \geq 0, \text{ and } i > 0 \text{ if } j \neq 0 \right\}$$
$$= k[x, xy, xy^2, \ldots, xy^n, \ldots].$$

It is clear that $(x, xy, xy^2, \ldots)$ is a maximal ideal, and is not finitely generated. (It looks as if it should be generated by x, but, of course, $y, y^2, \ldots$ are not elements of the ring.) Thus A_1 is not Noetherian.

(3) A rather similar example is the ring A_2 of polynomials in x, y, y^{-1} of the form $a(x, y) + xb(x, y, y^{-1})$; that is,

$$A_2 = \left\{ f(x, y) = \sum a_{ij} x^i y^j \mid i \geq 0, \text{ and } j \geq 0 \text{ if } i = 0 \right\}$$
$$= k[x, y, x/y, x/y^2, \ldots, x/y^n, \ldots].$$

In this ring $x = (x/y) \cdot y$, and $x/y = (x/y^2) \cdot y$, etc., so that the element x does not have a factorisation into irreducibles, and

$$(x) \subset (x/y) \subset (x/y^2) \subset \cdots$$

is an infinite ascending chain.

(4) Consider $0 \in \mathbb{R}$; a *function germ* at 0 is an equivalence class of continuous functions $f: U \to \mathbb{R}$ defined on an arbitrarily small neighbourhood U of 0 in $\mathbb{R}$; two germs $f: U \to \mathbb{R}$ and $g: V \to \mathbb{R}$ are equal by definition if and only if there exists a neighbourhood W of 0 with $W \subset U \cap V$ such that $f_{|W} = g_{|W}$. Function germs form a ring $\mathcal{E}_0$. This is a local ring, since the set of all function germs f such that $f(0) = 0$ is a maximal ideal m_0 (the kernel of evaluation at 0), and any function germ $f \notin m_0$ is invertible. Compare the ring of analytic function germs discussed in 1.15.

However, the ideal m_0 is not finitely generated. Indeed, suppose that f_i are any finite number of function germs vanishing at 0; if $g = \sum a_i f_i$ with a_i continuous functions of x, then we obviously get a bound $|g(x)| < \text{const.} \times \max |f_i(x)|$ as $x \to 0$. But, of course, there are functions g vanishing at 0, but vanishing much more slowly than $\max |f_i(x)|$. For example, set $G(x) = \max\{|x|, |f_i(x)|\}$ and $g(x) = \sqrt{G(x)}$; then

$$g(x)/ \max\{|x|, |f_i(x)|\} \to \infty \quad \text{as } x \to 0.$$

3.4 Noetherian modules

Definition An A-module M is *Noetherian* if the submodules of M have the a.c.c., that is, any increasing chain $M_1 \subset M_2 \subset \cdots \subset M_k \subset \cdots$ of submodules eventually stops. Just as before, it is equivalent to say that any nonempty set of submodules of M has a maximal element, or that every submodule of M is finite.

Proposition *Let* $0 \to L \xrightarrow{\alpha} M \xrightarrow{\beta} N \to 0$ *be a s.e.s. of A-modules. Then*

$$M \text{ is Noetherian} \iff L \text{ and } N \text{ are.}$$

Proof $\implies$ is easy, because ascending chains of submodules in L and N correspond one-to-one to certain ascending chains in M.

$\impliedby$ Suppose $M_1 \subset M_2 \subset \cdots \subset M_k \subset \cdots$ is an increasing chain of submodules of M; then identifying $\alpha(L)$ with L and taking intersection gives a chain

$$L \cap M_1 \subset L \cap M_2 \subset \cdots \subset L \cap M_k \subset \cdots$$

of submodules of L, and applying β gives a chain

$$\beta(M_1) \subset \beta(M_2) \subset \cdots \subset \beta(M_k) \subset \cdots$$

of submodules of N. Each of these two chains eventually stops, by the assumption on L and N, so that I need only prove the following statement:

Lemma *For submodules,* $M_1 \subset M_2 \subset M$,

$$L \cap M_1 = L \cap M_2 \quad and \quad \beta(M_1) = \beta(M_2) \implies M_1 = M_2.$$

Proof Indeed, if $m \in M_2$ then $\beta(m) \in \beta(M_1) = \beta(M_2)$, so that there is an $n \in M_1$ such that $\beta(m) = \beta(n)$. Then $\beta(m - n) = 0$, so that $m - n \in M_2 \cap \ker \beta = M_1 \cap \ker \beta$. Hence $m \in M_1$. Q.E.D.

3.5 Properties of Noetherian modules

Corollary

(i) *If M_i are Noetherian modules (for $i = 1, \ldots, r$) then $\bigoplus_{i=1}^{r} M_i$ is Noetherian.*

(ii) *If A is a Noetherian ring, then an A-module M is Noetherian if and only if it is finite over A.*

(iii) *If A is a Noetherian ring, M a finite A-module, then any submodule $N \subset M$ is again finite.*

(iv) *If A is a Noetherian ring and $\varphi \colon A \to B$ a ring homomorphism such that B is a finite A-module, then B is a Noetherian ring.*

Proof (i) A direct sum $M_1 \oplus M_2$ is a particular case of an exact sequence (compare Proposition 2.10), so that Proposition 3.4 proves (i) when $r = 2$. The case $r > 2$ follows by an easy induction.

(ii) If M is finite then there is a surjective homomorphism $A^r \to M \to 0$ for some r, so that M is a quotient $M \cong A^r/N$ for some submodule $N \subset A^r$; now A^r is a Noetherian module by (i), so M Noetherian follows by the implication $\implies$ of Proposition 3.4. Conversely, M Noetherian obviously implies M finite.

(iii) This just uses the previous implications: M finite and A Noetherian implies that M is Noetherian, so that N is Noetherian, which implies that N is a finite A-module.

(iv) B is Noetherian as an A-module; but ideals of B are submodules of B as an A-submodule, so that B is a Noetherian ring.   Q.E.D.

3.6 The Hilbert basis theorem

The following result provides many examples of Noetherian rings, and is the main motivation behind the use of the a.c.c. in commutative algebra. Note that in Hilbert's day, a "basis" of a module meant simply a family of generators.

Theorem   (Hilbert basis theorem) *If A is a Noetherian ring then so is the polynomial ring $A[X]$.*

Proof I prove that any ideal $I \subset A[X]$ is finitely generated. For this, define auxiliary sets $J_n \subset A$ by

$$J_n = \big\{ a \in A \mid \exists f \in I \text{ such that } f = aX^n + b_{n-1}X^{n-1} + \cdots + b_0 \big\}.$$

In other words, J_n is the set of leading coefficients of elements of I of degree n. Then it is easy to check that J_n is an ideal (using the fact that I is an ideal), and that $J_n \subset J_{n+1}$ (because for $f \in I$, also $Xf \in I$), and therefore

$$J_1 \subset J_2 \subset \cdots \subset J_k \subset \cdots$$

is an increasing chain of ideals. Using the assumption that A is Noetherian, I deduce that $J_n = J_{n+1} = \cdots$ for some n.

For each $m \leq n$, the ideal $J_m \subset A$ is finitely generated, say $J_m = (a_{m,1}, \ldots, a_{m,r_m})$; and by definition of J_m, for each $a_{m,j}$ with $1 \leq j \leq r_m$ there is a polynomial $f_{m,j} \in I$ of degree m having the leading coefficient

$a_{m,j}$. This allows me to write down a finite set

$$\{f_{m,j}\}_{\substack{m \leq n \\ 1 \leq j \leq r_m}} \qquad (*)$$

of elements of I.

I now claim that the finite set $(*)$ generates I. Indeed, for any polynomial $f \in I$, if f has degree m then its leading coefficient a is in J_m, hence if $m \geq n$, then $a \in J_m = J_n$, so that $a = \sum b_i a_{n,i}$ with $b_i \in A$ and $f - \sum b_i X^{m-n} f_{n,i}$ has degree $< m$; similarly, if $m \leq n$, then $a \in J_m$, so that $a = \sum b_i a_{m,i}$ with $b_i \in A$ and $f - \sum b_i f_{m,i}$ has degree $< m$. By induction on m it follows that f can be written as a linear combination of the finitely many elements $(*)$. This proves that any ideal of $A[X]$ is finitely generated.   Q.E.D.

Corollary *If A is a Noetherian ring, and $\varphi \colon A \to B$ a ring homomorphism such that B is a finitely generated extension ring of $\varphi(A)$, then B is Noetherian. In particular, any finitely generated algebra over $\mathbb{Z}$ or over a field k is Noetherian.*

Proof The assumption is that B is a quotient of a polynomial ring, $B \cong A[X_1, \ldots, X_n]/I$ for some ideal I. Now by Theorem 3.6 and an obvious induction, A Noetherian implies that so is $A[X_1, \ldots, X_n]$, and by Corollary 3.5, (iv), $A[X_1, \ldots, X_n]$ is Noetherian implies that so is $A[X_1, \ldots, X_n]/I$.   Q.E.D.

Exercises to Chapter 3

3.1   A subring of a Noetherian ring is Noetherian, true or false?

3.2   Let k be a field and $A \supset k$ a ring which is finite dimensional as a k-vector space; prove that A is Noetherian and Artinian (see 3.2 for the definition).

3.3   Let A be a ring and $I_1, \ldots, I_k$ ideals such that each A/I_i is a Noetherian ring. Prove that $\bigoplus A/I_i$ is a Noetherian A-module, and deduce that if $\bigcap I_i = 0$ then A is also Noetherian.

3.4   Prove that if A is a Noetherian ring and M a finite A-module, then there exists an exact sequence $A^q \xrightarrow{\alpha} A^p \xrightarrow{\beta} M \to 0$. That is, M has a presentation as an A-module in terms of finitely many generators and relations.

3.5   Let $0 \to L \xrightarrow{\alpha} M \xrightarrow{\beta} N \to 0$ be an exact sequence, and M_1, M_2 two submodules of M; decide whether the following implication holds:

$$\beta(M_1) = \beta(M_2) \text{ and } \alpha^{-1}(M_1) = \alpha^{-1}(M_2) \implies M_1 = M_2.$$

3.6   If A is a Noetherian ring, prove that any surjective ring homomorphism $\varphi \colon A \to A$ is also injective. [Hint: consider the chain of ideals $\ker \varphi \subset \ker \varphi^2 \subset \cdots$. If you can't do it from this hint, compare the proof of Lemma 7.11 and Ex. 7.5.]

3.7   Consider the $\mathbb{Z}$-module $M = (\mathbb{Z}[1/p])/\mathbb{Z}$; prove that any submodule $N \subset M$ is either finite (as a set) or the whole of M. Deduce that M is an Artinian $\mathbb{Z}$-module; prove that it is not finitely generated (so not Noetherian).

3.8   Let A be an Artinian integral domain (see 3.2 for the definition). Prove that A is a field. [Hint: for $f \in A$, the d.c.c. applied to $(f) \supset (f^2) \supset \cdots \supset (f^k)$ gives a relation $f^k = af^{k+1}$.] Deduce that every prime ideal of an Artinian ring is maximal.

3.9   Let A, m be an Artinian local ring (see 1.13 and 3.2). Prove that the maximal ideal m is nilpotent. More precisely, the d.c.c. gives that $m^k = m^{k+1}$ for some $k > 0$. Prove that $m^k = 0$. [Hint: suppose by contradiction that $m^k \neq 0$; let I be minimal among the ideals of A with $I \cdot m^k \neq 0$. Prove that $I = (x)$ is principal, then apply Nakayama's lemma 2.8.2 to it. Compare [M], Chapter 1, Theorem 3.2.]

3.10   The argument given in 3.1 to prove that the a.c.c. on a partially ordered set Σ implies the maximal condition involved making a countable number of choices, so it used implicitly a form of the axiom of choice. Write out a formal argument using Zorn's lemma. [Hint: Let Λ be the set of finite chains $\sigma : s_1 < s_2 < \cdots$ of Σ, ordered by inclusion (that is, $\sigma \leq \sigma'$ holds for two chains σ, σ' if σ is an initial segment of σ'). Use the a.c.c. to prove that a totally ordered subset of Λ has a maximal element.] An alternative proof based on the axiom of choice is given in [M], p. 14.

3.11   (Harder) Prove that if A is a Noetherian ring then so is the formal power series ring $A[\![X]\!]$. [Hint: as in the proof of the Hilbert basis theorem, consider an ideal $I \subset A[\![X]\!]$, and auxiliary ideals

$$J_n = \{a \in A \mid \exists f \in I \text{ such that } f = aX^n + b_{n+1}X^{n+1} + \cdots\};$$

show that the J_n form an increasing chain of ideals of A, and using this, find a finite set of power series to generate I. Compare [M], Chapter 1, Theorem 3.3.]

3.12   (Harder) Prove I. S. Cohen's theorem that if all the prime ideals of a ring A are finitely generated, then A is Noetherian. [Hint: the idea is simple: consider the set of all ideals which are not finitely generated; then Step 1. By Zorn's lemma, this has a maximal element P, and Step 2. P must be prime. See [M], Chapter 1, Theorem 3.4.]

4

Finite ring extensions and Noether normalisation

As you know, if $k \subset K$ is a field extension, an element $y \in K$ is *algebraic* over k if it satisfies an algebraic dependence relation

$$f(y) = a_n y^n + \cdots + a_1 y + a_0 = 0,$$

with some nonzero polynomial $f \in k[Y]$; without loss of generality, $a_n \neq 0$. Over a field, it costs nothing to divide the relation through by a_n, so we almost always just assume that $a_n = 1$, say that f is *monic*, and think no more about it. Over a more general ring A, however, you can't just divide by a_n. A relation

$$f(y) = y^n + a_{n-1} y^{n-1} + \cdots + a_1 y + a_0 = 0$$

with a monic polynomial $f \in A[Y]$ is called an *integral* dependence relation for y; the distinction between algebraic dependence and integral dependence is crucial. This notion leads at once to the ring of integers O_K of a number field K, which is the basic definition of algebraic number theory.

The close relation between integral dependence and finiteness conditions in a ring extension $A \subset B$ gives a convenient algebraic method for handling integral dependence: for example, it proves that if $y_1, y_2 \in B$ are integral over A, then so are $y_1 + y_2$ and $y_1 y_2$, without the need for calculating their integral dependence relations explicitly. These finiteness conditions lead naturally to Noether normalisation, which will be the key to the Nullstellensatz in the following chapter. At the same time, I give a brief introductory discussion in 4.5 of the relation between normal (integrally closed) rings and nonsingular algebraic curves, which is an ultimate aim of the final chapters of this book.

58

4.1 Finite and integral A-algebras

Let A be a ring; an A-*algebra* B is by definition a ring B with a given ring homomorphism $\varphi \colon A \to B$. The point is that B is then an A-module, with the multiplication defined by $\varphi(a) \cdot b$. An important case of this is when $A \subset B$, when B is also called an *extension ring* of A; we can usually reduce to this case by writing $\varphi(A) = A' \subset B$.

Definition Let B be an A-algebra.

(i) B is a *finite* A-algebra (or is *finite over* A) if it is finite as an A-module.

(ii) An element $y \in B$ is *integral* over A if there exists a monic polynomial $f(Y) = Y^n + a_{n-1}Y^{n-1} + \cdots + a_0 \in A'[Y]$ such that

$$f(y) = y^n + a_{n-1}y^{n-1} + \cdots + a_0 = 0. \tag{1}$$

The algebra B is *integral* over A if every $b \in B$ is integral. (The terminology integral domains and integral extension are unrelated.)

In terms of B viewed as an A-module, the integral dependence relation (1) is a linear relation

$$1_A \cdot y^n + a_{n-1} \cdot y^{n-1} + \cdots + a_0 \cdot 1_B = 0 \tag{2}$$

between the powers of y, with coefficients in A, and such that the highest power y^n has coefficient 1.

Examples

(i) $\mathbb{Z}[1/3]$ is not integral over $\mathbb{Z}$, and for $f \in k[X]$ an irreducible polynomial, $k[X][1/f]$ is not integral over $k[X]$; more generally, by Ex. 0.7, if A is a UFD with field of fractions K then no element of $K \setminus A$ is integral over A. These extensions are also not finite by Ex. 2.1.

(ii) $k[X^2] \subset k[X]$ is an integral extension.

(iii) $\mathbb{Z} \subset \mathbb{Z}[\tau]$ is integral, where $\tau = (1 + \sqrt{5})/2$ is the "golden ratio", because $\tau^2 - \tau - 1 = 0$; on the other hand $\alpha = (1 + \sqrt{3})/2$ is not integral over $\mathbb{Z}$ (because $\alpha^2 - \alpha = 1/2$).

4.2 Finite versus integral

Proposition *Let* $\varphi\colon A \to B$ *be an A-algebra, and* $y \in B$; *then three equivalent conditions:*

   (i) y *is integral over* A.

   (ii) *The subring* $A'[y] \subset B$ *generated by* $A' = \varphi(A)$ *and* y *is finite over* A.

   (iii) *There exists an A-subalgebra* $C \subset B$ *such that* $A'[y] \subset C$ *and* C *is finite over* A.

Proof (i) $\Longrightarrow$ (ii) Exactly as in field theory, if y satisfies a relation (1) then $A'[y]$ is generated by $1, y, \ldots, y^{n-1}$. Indeed, given any element $b_m y^m + \cdots + b_1 y + b_0 \in A'[y]$, if $m \geq n$ then subtracting $b_m y^{m-n}$ times the relation (1) kills the leading term.

(ii) $\Longrightarrow$ (iii) is trivial: just take $C = A'[y]$.

(iii) $\Longrightarrow$ (i) Consider the A-module homomorphism $f = \mu_y\colon C \to C$ defined by multiplication by y. Then C is a finite A-module, so that I can apply the determinant trick 2.7 to f. This gives a relation

$$f^n + a_{n-1} \cdot f^{n-1} + \cdots + a_0 = 0 \qquad \text{(as endomorphisms } C \to C\text{)}.$$

However, $f^i(1) = y \cdot y \cdots y \cdot 1$ (i times) $= y^i$, so that this gives the required relation.   Q.E.D.

Remark To prove (iii) $\Longrightarrow$ (i) for a field extension $A = k \subset C$, we just say that $1, y, y^2, \ldots$ are infinitely many elements of the finite dimensional vector space C, and deduce a linear dependence relation between them, which without loss of generality is a monic polynomial relation for y over k. This direct line of argument does not work at this level of generality (see below for some attempts), which is why I used the determinant trick. In fact, this is the main use of the determinant trick in commutative algebra.

Alternative proof of (ii) $\Longrightarrow$ (i) Because $A'[y] = \sum_{i=0}^{\infty} Ay^i \subset B$, if it is finite over A then obviously $A'[y] = \sum_{i=0}^{n} Ay^i$ for some n. Then $y^{n+1} \in \sum_{i=0}^{n} Ay^i$ gives the required integral dependence relation for y.

Alternative proof of (iii) $\Longrightarrow$ (i), *assuming A is Noetherian* If I add the assumption that A is Noetherian, then C is a Noetherian module (by

Corollary 3.5); then $\sum_{i=0}^{k} Ay^i$ form an ascending chain of submodules of C, therefore $y^{n+1} \in \sum_{i=0}^{n} Ay^i$ for some n, and I conclude as before.

4.3 Tower laws

For simplicity of notation, from now on I only deal with extension rings $A \subset B$; it will usually be clear whether some fact actually holds more generally for any A-algebra B.

Proposition *Let B be an A-algebra.*

(a) *If $A \subset B \subset C$ are extension rings such that C is a finite B-algebra and B a finite A-algebra, then C is finite over A.*

(b) *If $y_1, \ldots, y_m \in B$ are integral over A then $A[y_1, \ldots, y_m]$ is finite over A; in particular, every $f \in A[y_1, \ldots, y_m]$ is integral over A.*

(c) *If $A \subset B \subset C$ with C integral over B and B integral over A then C is integral over A.*

(d) *The subset $\widetilde{A} = \{y \in B \mid y \text{ is integral over } A\} \subset B$ is a subring of B; moreover, if $y \in B$ is integral over $\widetilde{A}$ then $y \in \widetilde{A}$, so that $\widetilde{\widetilde{A}} = \widetilde{A}$.*

Proof (a) is proved exactly as for field extensions, and is left as an easy exercise.

(b) $A \to A[y_1]$ is finite by Proposition 4.2, (i) $\implies$ (ii). The result for $A[y_1, \ldots, y_m]$ follows by induction using (a).

(c) Let $z \in C$; then z satisfies a relation $z^n + b_{n-1}z^{n-1} + \cdots + b_0 = 0$, with $b_0, \ldots, b_{n-1} \in B$. Now each b_i is integral over A, so each of the steps $A \subset A[b_0, \ldots, b_{n-1}] \subset A[b_0, \ldots, b_{n-1}, z]$ is finite, and thus z belongs to an intermediate algebra $A \subset A[b_0, \ldots, b_{n-1}, z] \subset C$ which is finite over A; this is condition 4.2, (iii).

For the first part of (d), note that if $\alpha, \beta \in \widetilde{A}$ then $\alpha \pm \beta$ and $\alpha\beta$ are both elements of $A[\alpha, \beta]$, which by (b) is finite over A. The last part is clear from (c).   Q.E.D.

4.4 Integral closure

Definition The ring $\widetilde{A}$ appearing in Proposition 4.3, (d) is the *integral closure* of A in B; if $A = \widetilde{A}$ then A is *integrally closed* in B. An integral domain A is *normal* if it is integrally closed in its field of fractions, that is, if $A = \widetilde{A} \subset K = \operatorname{Frac} A$; for any integral domain A, the integral closure of A in its field of fractions $K = \operatorname{Frac} A$ is also called the *normalisation* of A.

An important particular case of this definition is the ring of integers in algebraic number theory: a *number field* is a finite field extension $\mathbb{Q} \subset K$ of the rational field (that is, $[K : \mathbb{Q}] = \dim_{\mathbb{Q}} K < \infty$). The *ring of integers* O_K of K is the ring $\{y \in K \mid y$ is integral over $\mathbb{Z}\}$, the integral closure of $\mathbb{Z}$ in K.

Examples

(i) A UFD is normal, as you proved in Ex. 0.7.

(ii) Consider the quadratic number field $\mathbb{Q} \subset \mathbb{Q}(\sqrt{n})$, where n is a square-free integer, and the subring $\mathbb{Z} \subset \mathbb{Q}[\sqrt{n}]$. Set

$$\begin{cases} \alpha = (1 + \sqrt{n})/2 & \text{if } n \equiv 1 \bmod 4 \\ \alpha = \sqrt{n} & \text{if } n \equiv 2 \text{ or } 3 \bmod 4; \end{cases}$$

then it is not hard to see that the integral closure of $\mathbb{Z}$ in $\mathbb{Q}(\sqrt{n})$ is $\mathbb{Z}[\alpha]$.

(iii) Consider the ring $A = k[X,Y]/(Y^2 - X^3)$, and write $x, y \in A$ for the classes of X, Y; then A is not normal: it is not hard to see that it has field of fractions $\operatorname{Frac} A = k(t)$, where $t = y/x$, and $x = t^2$, $y = t^3$; either of these relations show that t is integral over A, but obviously $t \notin A$. Also, $k[t]$ is normal (because it's a UFD), so it is the integral closure of A in $k(t)$. Compare 0.7 for more discussion.

4.5 Preview: nonsingularity and normal rings

Example (iii) hints that the singularity of the plane curve $Y^2 = X^3$ at the origin is somehow related to the nonnormality of the ring $A = k[X,Y]/(Y^2 - X^3)$. The normalisation $A \subset \tilde{A} = k[t]$ corresponds to a parametrisation $t \mapsto (t^2, t^3)$ of the curve $Y^2 = X^3$. I don't have time to discuss this in detail here, but I state the following results, which are proved in Chapter 8 (in slightly different language).

Let k be an algebraically closed field, $f \in k[X,Y]$ an irreducible polynomial, and $C \subset k^2$ the plane curve defined by $f(X,Y) = 0$. Then $A = k[X,Y]/(f) = k[C]$ is the ring of polynomial functions on C. We say that C is *nonsingular* if $(\partial f/\partial x, \partial f/\partial y)(P) \neq (0,0)$ at every point $P \in C$ (so that C has a well-defined tangent line at P).

Fact 1 *A is normal $\iff$ C is nonsingular.*

Fact 2 *Let $\tilde{A} \subset K = \text{Frac}\, A$ be the normalisation of A. Then there is a nonsingular algebraic curve $\tilde{C}$ (contained in some k^n) with the property that the normalisation $\tilde{A} = k[\tilde{C}]$ is the ring of polynomial functions on $\tilde{C}$, and the inclusion $A \subset \tilde{A}$ corresponds to a polynomial map $\tilde{C} \to C$ between algebraic curves. That is, the normalisation of A corresponds to a resolution of singularities; for example, the above parametrisation $k \to C$ given by $t \mapsto (t^2, t^3)$.*

Thus, amazingly, the same algebraic apparatus gives the ring of integers of a number field and the nonsingularity of algebraic curves.

4.6 Noether normalisation

For the rest of this section k is a field, and A a k-algebra.

Definition Elements $y_1, \ldots, y_n \in A$ are *algebraically independent* over k if the natural surjection $k[Y_1, \ldots, Y_n] \to k[y_1, \ldots, y_n]$ is an isomorphism, where the left-hand side is the polynomial ring. This just means that there are no nonzero polynomial relations $F(y_1, \ldots, y_n) = 0$ with coefficients in k.

Recall that a k-algebra A is *finitely generated* (f.g.) over k if $A = k[y_1, \ldots, y_n]$ for some finite set $y_1, \ldots, y_n$.

Theorem (Noether normalisation lemma) *Let k be a field, and A a f.g. k-algebra; then there exist elements $z_1, \ldots, z_m \in A$ such that*

(i) *$z_1, \ldots, z_m$ are algebraically independent over k;*
(ii) *A is finite over $B = k[z_1, \ldots, z_m]$.*

That is, a f.g. extension $k \subset A$ can be written as a composite

$$k \subset B = k[z_1, \ldots, z_m] \subset A,$$

where $k \subset B$ is a polynomial extension, and $B \subset A$ is finite. This is a rough structure theorem for finitely generated k-algebras.

Main Claim *Suppose that $A = k[y_1, \ldots, y_n]$, and that*

$$0 \neq F \in k[Y_1, \ldots, Y_n]$$

is such that $F(y_1, \ldots, y_n) = 0$. Then there exist $y_1^*, \ldots, y_{n-1}^* \in A$ such that y_n is integral over $A^* = k[y_1^*, \ldots, y_{n-1}^*]$ and $A = A^*[y_n]$.

Proof of Theorem, assuming the claim This is an easy induction on the number n of generators of A: if $n = 0$, there's nothing to prove. If $n > 0$ and $y_1, \ldots, y_n$ are algebraically independent over k, then again there's nothing to prove. So suppose that $y_1, \ldots, y_n$ are algebraically dependent over k, and let $0 \neq F \in k[Y_1, \ldots, Y_n]$ be such that $F(y_1, \ldots, y_n) = 0$; then, by the claim, there exist $y_1^*, \ldots, y_{n-1}^* \in A$ such that y_n is integral over $A^* = k[y_1^*, \ldots, y_{n-1}^*]$ and $A = A^*[y_n]$.

Now by the inductive hypothesis applied to A^*, there exist elements $z_1, \ldots, z_m \in A^*$ that are algebraically independent over k and with A^* finite over $B = k[z_1, \ldots, z_m]$. Now since y_n is integral over A^*, it follows that $A^*[y_n]$ is finite over A^*, so that each step $B \subset A^* \subset A^*[y_n] = A$ is finite, and A is finite over B, as required.   Q.E.D.

4.7 Proof of Claim

Let $r_1, \ldots, r_{n-1} \geq 1$ be integers (to be specified later), and set

$$y_i^* = y_i - y_n^{r_i} \qquad \text{for } i = 1, \ldots, n-1.$$

Then define $G \in A$ by $G(y_1^*, \ldots, y_{n-1}^*, y_n) = F(y_i^* + y_n^{r_i}, y_n) = 0$. I view this as a relation for y_n over $k[y_1^*, \ldots, y_{n-1}^*]$, and endeavour to prove that it is an integral dependence relation for suitable choice of $r_1, \ldots, r_{n-1}$.

Since F is a polynomial in $y_1, \ldots, y_n$, I can write it as a sum of monomials

$$F = \sum_m a_m y^m = \sum_m a_m \prod y_i^{m_i},$$

where $m = (m_1, \ldots, m_n)$, and each $a_m \neq 0$. Therefore

$$G = \sum a_m \prod (y_i^* + y_n^{r_i})^{m_i}.$$

When I expand out, each summand $a_m \prod (y_i^* + y_n^{r_i})^{m_i}$ has a unique term of highest order in y_n, namely $a_m y_n^{\sum r_i m_i}$; now suppose that I can arrange that

$$m \neq m' \implies \sum r_i m_i \neq \sum r_i m_i'.$$

Then $\max\{\sum r_i m_i \mid m \text{ such that } a_m \neq 0\}$ is achieved in only one summand, so that there can be no cancellation, and the highest order term in y_n occurring in G is the term $a_m y_n^{\sum r_i m_i}$ coming from this; this is a_m times a pure power of y_n, so that G is the field element $a_m \neq 0$ times a monic polynomial in y_n. Therefore I only need to prove the following result.

Horrible lemma *Given a finite set $\{Y^m\}$ of monomials in $Y_1, \ldots, Y_n$, there exists a system of integers $w_1, \ldots, w_{n-1}, w_n = 1$ (called* weights*) such that the weight of a monomial*

$$w(Y^m) = \sum_{i=1}^{n} w_i m_i$$

distinguishes the monomial, that is,

$$m \neq m' \implies w(Y^m) \neq w(Y^{m'}).$$

Proof By induction on n, the case $n = 1$ being trivial. Write each monomial as $Y_1^{m_1} Y^{m'}$, with $m' = (m_2, \ldots, m_n)$. Then I get a finite set of monomials $\{Y^{m'}\}$ in the $n - 1$ variables $Y_2, \ldots, Y_n$, so that I can choose weights $w_2, \ldots, w_n$ to distinguish these. Now if I just choose $w_1 > \max w(Y^{m'})$, the weights $w_1, \ldots, w_n$ clearly distinguish the $\{Y^m\}$. In other words, I can force a lex ordering of monomials by weighting Y_1 much more heavily than $Y_2, \ldots, Y_n$, and Y_2 much more heavily than $Y_3, \ldots, Y_n$, etc.

This completes the proof of Theorem 4.6. Q.E.D.

4.8 Another proof of Noether normalisation

The proof of the main claim just given is due to Nagata in the 1950s. There is an alternative proof traditional in algebraic geometry, which works if the field k is infinite. Let $\alpha_1, \ldots, \alpha_{n-1} \in k$, and set

$$y_i^* = y_i - \alpha_i y_n \qquad \text{for } i = 1, \ldots, n - 1.$$

Then it's not hard to see that if k is infinite, the α_i can be chosen so that $F(y_i^* + \alpha_i y_n, y_n) = 0$ is a monic relation for y_n over $k[y_1^*, \ldots, y_{n-1}^*]$. See Ex. 4.11 for details, or any textbook on algebraic geometry, for example, [UAG], §3.

Example To illustrate the above proof, consider $A = k[X, Y]/(XY - 1)$. Then Y is algebraic over $k[X]$ (in fact, $Y = X^{-1} \in k(X)$), but not integral over $k[X]$. This corresponds to the fact that the hyperbola $XY = 1$ has the line $X = 0$ as an asymptotic line, so that its projection to the X-axis has a "missing" root over $X = 0$. If I take $X' = X - \varepsilon Y$ as the basic transcendental element of A instead of X, the relation becomes $(X' + \varepsilon Y)Y = 1$, which is monic in Y if $\varepsilon \neq 0$. This corresponds

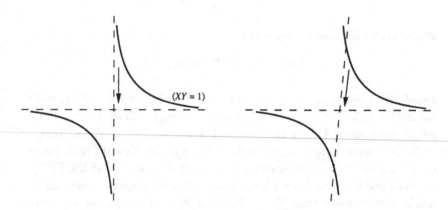

Figure 4.8. Projecting the hyperbola $XY = 1$

geometrically to tilting the hyperbola a little before projecting, so that it no longer has a vertical asymptotic line. See Figure 4.8.

The moral of either proof is that for any choice $y_1, \ldots, y_{n-1}$, we'll be fairly unlucky if y_n is not integral over $k[y_1, \ldots, y_{n-1}]$. Perturbing the choice of $y_1, \ldots, y_{n-1}$, I can guarantee to avoid the unlucky accident.

4.9 Field extensions

Proposition *Let $A \subset B$ be an integral extension of integral domains; then*

$$A \text{ is a field} \iff B \text{ is a field.}$$

Proof $\implies$ Let $0 \neq x \in B$; then there is a monic relation

$$x^n + a_{n-1}x^{n-1} + \cdots + a_0 = 0$$

with $a_i \in A$, and I can assume that $a_0 \neq 0$, since otherwise I could cancel x. Now A is a field, therefore

$$x^{-1} = -a_0^{-1}(x^{n-1} + a_{n-1}x^{n-2} + \cdots + a_2 x + a_1) \in B.$$

$\Longleftarrow$ Similarly, if B is a field and $0 \neq x \in A$ then $x^{-1} \in B$, and so x^{-1} is integral over A. So there is a relation of the form

$$(x^{-1})^n + a_{n-1}(x^{-1})^{n-1} + \cdots + a_0 = 0,$$

and therefore

$$x^{-1} = -a_{n-1} - a_{n-2}x - \cdots - x^{n-1}a_0 \in A. \quad \text{Q.E.D.}$$

4.10 The weak Nullstellensatz

Theorem (Weak Nullstellensatz) *Let k be a field, and K a k-algebra which*

(i) *is finitely generated as a k-algebra; and*
(ii) *is a field.*

Then K is algebraic over k, so that $k \subset K$ is a finite field extension; that is, $[K : k] < \infty$.

Proof By Noether normalisation, there exist elements $z_1, \ldots, z_m \in K$ which are algebraically independent, and such that K is finite over $A = k[z_1, \ldots, z_m]$. But now we're in the situation of Proposition 4.9: $A \subset K$ is integral, K is a field, so therefore A is a field. Because $z_1, \ldots, z_m \in K$ are algebraically independent, $A = k[z_1, \ldots, z_m]$ is a polynomial ring in m indeterminates, and this is a field only if $m = 0$, and K is finite over k itself. Q.E.D.

Exercises to Chapter 4

4.1   (a) $k[X^2] \subset k[X]$ is a finite extension, hence integral. Find the integral dependence relation for any $f \in k[X]$.

            (b) Find the integral dependence relation over $\mathbb{Z}$ satisfied by an element $a + b\tau \in \mathbb{Z}[\tau]$, where τ is the "golden ratio" (see 4.1).

4.2 $A \subset B$ is a ring extension. If $y, z \in B$ satisfy quadratic integral dependence relations $y^2 + ay + b = 0$ and $z^2 + cz + d = 0$ over A, find explicit integral dependence relations for $y + z$ and yz. [Hint: start by assuming $1/2 \in A$ and completing the square.]

4.3   If $y, z \in B$ are integral over a subring A, use elementary symmetric functions to prove that so are $y + z$ and yz. [Hint: there exists an extension ring in which the monic relations f, g for y, z split into linear factors. The coefficients of f, g are the elementary symmetric functions in the roots y_i, z_j. You have to prove that the elementary symmetric functions in $y_i + z_j$ can be expressed in terms of them.] Trying to calculate explicitly the monic relation for $y + z$ in any case with $\deg f, \deg g \geq 3$ will convince you of the advantage of the more abstract "finite versus integral" method.

4.4   Let n be a square-free integer, and consider the quadratic number field $k = \mathbb{Q}(\sqrt{n})$; prove that $\alpha = a + b\sqrt{n} \in k$ is integral over $\mathbb{Z}$ if and only if either $a, b \in \mathbb{Z}$ or $n \equiv 1 \bmod 4$ and $a \equiv b \equiv 1/2 \bmod \mathbb{Z}$.

4.5   Let $A = k[X, Y]/(Y^2 - X^2 - X^3)$. As in 4.4, Example (iii), prove that the normalisation of A is $k[t]$ where $t = Y/X$.

4.6   Let $A = k[X]$, and $f \in A$; then set $B = k[X, Y]/(Y^2 - f)$. Find the necessary and sufficient condition that f must satisfy in order for B to be an integral domain. Assume this holds, and write K for the field of fraction of B, that is $K = k(X)(\sqrt{f})$. For any $\alpha \in K$, write down a polynomial $h(T) \in A[T]$ such that $h(\alpha) = 0$. [Hint: Write $\alpha = a + b\sqrt{f}$ with $a, b \in K(X)$.]

     Show that B is normal if and only if f has no square factors. If B is not normal, show how to obtain its normalisation in terms of f.

4.7   Find the normalisation of the rings (i) $k[X, Y]/(Y^3 - X^5)$, and (ii) $k[X, Y]/(Y^5 - X^{19})$.

4.8   Let $n \in \mathbb{Z}$ be a number not divisible by any p^3. Find the normalisation of $\mathbb{Z}[\sqrt[3]{n}]$. [Hint: suppose $n = l^2 m$; then the field $\mathbb{Q}(\sqrt[3]{n})$ also contains $\sqrt[3]{lm^2}$. Write any element of $\mathbb{Q}(\sqrt[3]{n})$ in the form $\alpha = a + b\sqrt[3]{n} + c\sqrt[3]{lm^2}$ with $a, b, c \in \mathbb{Q}$, and calculate its minimal polynomial over $\mathbb{Q}$.]

4.9   Let k be an arbitrary field, and

$$A = k[X, Y, Z]/(X^2 - Y^3 - 1, XZ - 1);$$

find $\alpha, \beta \in k$ such that A is integral over $B = k[X + \alpha Y + \beta Z]$, and for your choice write down a set of generators of A as a B-module.

4.10   Suppose $k = \mathbb{F}_q$ is a finite field with q elements; give an example of $f \in k[X, Y]$ such that for every $\alpha \in k$, the ring $A = k[X, Y]/(f)$ is not finitely generated as a module over $B_\alpha = k[X - \alpha Y]$. (In other words, the alternative proof of

4.8 does not work in this case.) [Hint: use the fact that $\alpha^q = \alpha$ for all $\alpha \in k$.]

4.11   Assuming that the field k is infinite, complete the alternative proof of the Main Claim 4.6 discussed in 4.8: let $\alpha_1, \ldots, \alpha_{n-1} \in k$, and set $y_i^* = y_i - \alpha_i y_n$ for $i = 1, \ldots, n-1$. Then in A,

$$G(y_1^*, \ldots, y_{n-1}^*, y_n) = F(y_i^* + \alpha_i y_n, y_n) = 0.$$

If F has degree d, then the coefficient of y_n^d in G is

$$G = F_*(\alpha_1, \ldots, \alpha_{n-1}, 1) \cdot y_n^d + \text{lower order terms in } y_n,$$

where F_* is the homogeneous piece of F of degree d. Prove that for an infinite field k, and any homogeneous $F_* \neq 0$, there exist elements $\alpha_1, \ldots, \alpha_{n-1} \in k$ such that $F_*(\alpha_1, \ldots, \alpha_{n-1}, 1) \neq 0$. Then for this choice, the relation $G(y_1^*, \ldots, y_{n-1}^*, y_n) = 0$ is an integral dependence relation for y_n over $A' = k[y_1^*, \ldots, y_{n-1}^*]$.

4.12   (i) Let $A \subset B$ be a finite ring extension and $P \subset A$ a prime ideal; prove that there exists at least one prime ideal Q of B with $Q \cap A = P$. [Hint: use Nakayama's lemma to deduce that $PB \neq B$.]

    (ii) If k is a field and $k \subset B$ a finite ring extension, prove that B has only finitely many prime ideals Q. [Hint: prove that, if P is a prime ideal, then $k(P) = B/P$ is a field extension of k. Let $B = k[x_1, \ldots, x_n]$, and let $f_i(x_i) = 0$ be relations for x_i over k. Then there exists a field extension $k \subset K$ in which f_i has a full set of roots α_{ij}. Every field extension $k \subset k(P)$ (up to isomorphism) is obtained by sending each x_i to some α_{ij}, which gives only finitely many choices.]

    (iii) Look forward to Ex. 6.7 to see that in the general situation of (i), there are only finitely many prime ideals Q of B with $Q \cap A = P$.

5

The Nullstellensatz and the geometry of Spec A

This chapter is the core of the book. In it, I introduce the relation between varieties $V \subset k^n$ and f.g. k-algebras $A = k[X_1, \ldots, X_n]/I$, and prove the Nullstellensatz. The bridge between rings and spaces which is the main theme of the book takes a very concrete form when $A = k[V]$ is the coordinate ring of a variety $V \subset k^n$ over an algebraically closed field k. In addition, I discuss how this ideal case can be used to illuminate the study of quite general rings.

5.1 Weak Nullstellensatz

Let k be a field, and consider the polynomial ring $k[X_1, \ldots, X_n]$ in n variables. For any maximal ideal m, the residue field $K = k[X_1, \ldots, X_n]/m$ satisfies the conditions of the weak Nullstellensatz 4.10, so that K is a finite algebraic extension of k.

If I write $\alpha_1, \ldots, \alpha_n \in K$ for the residue classes of $X_1, \ldots, X_n$, the quotient homomorphism $k[X_1, \ldots, X_n] \to K = k[X_1, \ldots, X_n]/m$ is given by $f(X_1, \ldots, X_n) \mapsto f(\alpha_1, \ldots, \alpha_n)$, that is, evaluating polynomials f at the point $\alpha_1, \ldots, \alpha_n \in K^n$.

5.2 Maximal ideals of $k[X_1, \ldots, X_n]$ and points of k^n

Corollary *Suppose in addition that k is algebraically closed. Then every maximal ideal of $A = k[X_1, \ldots, X_n]$ is of the form*

$$m = (X_1 - a_1, \ldots, X_n - a_n) \qquad \text{for some } a_1, \ldots, a_n \in k;$$

the map $k[X_1, \ldots, X_n] \to k[X_1, \ldots, X_n]/m = k$ is just the natural "evaluation" map $f(X_1, \ldots, X_n) \mapsto f(a_1, \ldots, a_n)$.

Hence there is a natural one-to-one correspondence

$$k^n \leftrightarrow \text{m-Spec}\, A \qquad \textit{given by } (a_1, \ldots, a_n) \leftrightarrow (X_1 - a_1, \ldots, X_n - a_n).$$

As an easy exercise, think through what this means in the case $n = 1$. Also, work out how the case $n = 2$ of the theorem follows from the worked example 1.5.

Proof Let $K = k[X_1, \ldots, X_n]/m$; then $k \subset K$ is an algebraic extension. But since k is algebraically closed, $k = K$. In other words, the copy of k contained as a subring in the k-algebra $A = k[X_1, \ldots, X_n]$ maps isomorphically to the quotient $k[X_1, \ldots, X_n]/m$.

If $X_i \in A$ maps to a_i in the quotient $K = k$ then the same element $a_i \in k \subset A$ also maps to a_i, so $X_i - a_i \in m$ for $i = 1, \ldots, n$, and therefore $m \supset (X_1 - a_1, \ldots, X_n - a_n)$. However, $(X_1 - a_1, \ldots, X_n - a_n)$ is obviously a maximal ideal of $k[X_1, \ldots, X_n]$, because (for example) it is a k-vector subspace of codimension 1. (You've already seen a better proof in Ex. 1.15.) Q.E.D.

5.3 Definition of a variety

Let k be a field. A *variety* $V \subset k^n$ is a subset of the form

$$V = V(J) = \big\{ P = (a_1, \ldots, a_n) \in k^n \mid f(P) = 0 \text{ for all } f \in J \big\},$$

where $J \subset k[X_1, \ldots, X_n]$ is an ideal. Note that $J = (f_1, \ldots, f_m)$ is finitely generated, so that a variety V is defined by

$$f_1(P) = \cdots = f_m(P) = 0,$$

that is, it is a subset $V \subset k^n$ defined as the simultaneous solutions of a number of polynomial equations.

Proposition *Suppose that k is an algebraically closed field and that $A = k[x_1, \ldots, x_n]$ is a finitely generated k-algebra of the form $A = k[X_1, \ldots, X_n]/J$, where J is an ideal of $k[X_1, \ldots, X_n]$. (Here I'm using the notation to mean that $x_i = X_i \bmod J$.)*

Then every maximal ideal of A is of the form $(x_1 - a_1, \ldots, x_n - a_n)$ for some point $(a_1, \ldots, a_n) \in V(J)$. Therefore there is a one-to-one correspondence

$$V(J) \leftrightarrow \text{m-Spec}\, A \qquad \textit{given by } (a_1, \ldots, a_n) \leftrightarrow (x_1 - a_1, \ldots, x_n - a_n).$$

Proof The ideals of A are given by ideals of $k[X_1, \ldots, X_n]$ containing J. So every maximal ideal of A is of the form $(x_1 - a_1, \ldots, x_n - a_n)$ for some $a_1, \ldots, a_n$ such that

$$J \subset (X_1 - a_1, \ldots, X_n - a_n).$$

However, since $(X_1 - a_1, \ldots, X_n - a_n)$ is just the kernel of the evaluation map $f \mapsto f(a_1, \ldots, a_n)$, it follows that

$$J \subset (X_1 - a_1, \ldots, X_n - a_n) \iff f(a_1, \ldots, a_n) = 0 \text{ for every } f \in J,$$

that is, $(a_1, \ldots, a_n) \in V(J)$. Q.E.D.

5.4 Remark on algebraically nonclosed k

Throughout the discussion of varieties, I assume that k is algebraically closed for simplicity and to avoid appealing to definitions and results of Galois theory. It is not hard to give statements of the above results for any field. For example, for any field k, every maximal ideal of $k[X_1, \ldots, X_n]$ is given in the following way: take a finite algebraic extension $k \subset K$, and a point $(a_1, \ldots, a_n) \in K^n$; consider the ideal $(X_1 - a_1, \ldots, X_n - a_n)$ of $K[X_1, \ldots, X_n]$ and set

$$m = (X_1 - a_1, \ldots, X_n - a_n) \cap k[X_1, \ldots, X_n].$$

The effect of taking the intersection with $k[X_1, \ldots, X_n]$ is to consider $(a_1, \ldots, a_n) \in K^n$ up to conjugacy over k in the sense of Galois theory. See Ex. 1.16 and Exs. 5.6–7.

5.5 The correspondences V and I

A variety $X \subset k^n$ is by definition equal to $X = V(J)$, where J is an ideal of $k[X_1, \ldots, X_n]$; so V gives a map

$$\{\text{ideals of } k[X_1, \ldots, X_n]\} \xrightarrow{V} \{\text{subsets } X \text{ of } k^n\}.$$

There is a correspondence going the other way:

$$\{\text{subsets } X \text{ of } k^n\} \xrightarrow{I} \{\text{ideals of } k[X_1, \ldots, X_n]\},$$

defined by taking a subset $X \subset k^n$ into the ideal

$$I(X) = \{f \in k[X_1, \ldots, X_n] \mid f(P) = 0 \text{ for all } P \in X\}.$$

V and I satisfy a number of formal properties, such as reversing inclusions:

$$J \subset J' \implies V(J) \supset V(J') \quad \text{and} \quad X \subset Y \implies I(X) \supset I(Y).$$

Several others are discussed in 5.9 below and in the exercises. Tautologically, $X \subset V(I(X))$ for any subset X, and $X = V(I(X))$ if and only if X is a variety. Conversely, $J \subset I(V(J))$ for any ideal J.

5.6 The Nullstellensatz

Theorem (Nullstellensatz) *Let k be an algebraically closed field.*

(a) *If $J \subsetneq k[X_1, \ldots, X_n]$ then $V(J) \neq \emptyset$.*
(b) *$I(V(J)) = \operatorname{rad} J$; in other words, for $f \in k[X_1, \ldots, X_n]$,*

$$f(P) = 0 \text{ for all } P \in V \iff f^n \in J \text{ for some } n.$$

Remarks

(i) The name of the theorem "zeros theorem" comes from the statement (a): given a set of polynomials that generate a nontrivial ideal, they have at least one simultaneous zero. Note that the theorem is of course false if k is not algebraically closed: if $f \in k[X]$ is a polynomial of degree ≥ 2 with no roots in k then $(f) \neq k[X]$ but $V(f) = \emptyset$, so (a) fails; moreover $I(V(f)) = k[X]$, so (b) also fails.

(ii) Contrast the statement (b) with Corollary 1.12, which said that

$$\operatorname{rad} J = \bigcap_{\substack{P \in \operatorname{Spec} A \\ P \supset I}} P.$$

The Nullstellensatz is stronger than it in two respects: firstly (b) says that we can take the intersection just over maximal ideals of $k[X_1, \ldots, X_n]$; secondly we have a very nice description of the maximal ideals in terms of points $(a_1, \ldots, a_n)$ of n-dimensional space k^n. See 5.14 for more discussion. Corollary 1.12 could reasonably be called the *easy Nullstellensatz*.

Proof (a) This is easy, given Corollary 5.2. If J is a nontrivial ideal, it is contained in a maximal ideal $m = (X_1 - a_1, \ldots, X_n - a_n)$, and hence $P = (a_1, \ldots, a_n) \in V(J)$.

(b) is a consequence of (a), but requires the special trick invented by S. Rabinowitch in 1929; suppose $f \in k[X_1,\ldots,X_n]$ is such that $f(P) = 0$ for every point $P \in V(J)$. I introduce the auxiliary variable Y, and consider the ideal

$$J' = (J, fY - 1) \subset k[X_1,\ldots,X_n,Y].$$

Then a point of $V(J')$ is an $(n+1)$-tuple $(a_1,\ldots,a_n,b)$ of elements of k such that $(a_1,\ldots,a_n) \in V(J)$ and $bf(a_1,\ldots,a_n) = 1$; that is, $P \in V(J)$ and $f(P) \neq 0$. Therefore $V(J') = \emptyset$. By (a), this means that in $k[X_1,\ldots,X_n,Y]$, there is an identity

$$1 = \sum g_i h_i + g_0(fY - 1) \qquad \text{with } g_i \in k[X_1,\ldots,X_n,Y] \text{ and } h_i \in J.$$

If I multiply both sides by f^m, I can arrange that Y only appears in combination as fY, that is,

$$f^m = \sum G_i(X_1,\ldots,X_n,fY)h_i + G_0(X_1,\ldots,X_n,fY)(fY - 1).$$

This is an identity between polynomials, so it remains true if I substitute $fY = 1$. This gives $f^m \in J$, as required.   Q.E.D.

5.7 Irreducible varieties

Definition A variety $X \subset k^n$ is *irreducible* if it is nonempty and not the union of two proper subvarieties; that is, if

$$X = X_1 \cup X_2 \text{ for varieties } X_1, X_2 \implies X = X_1 \text{ or } X_2.$$

See Exs. 5.2–4 for examples.

Proposition *A variety X is irreducible if and only if $I(X)$ is prime.*

Proof Set $I = I(X)$; if I is not prime then let $f, g \in A \setminus I$ be such that $fg \in I$; now define new ideals $J_1 = (I, f)$ and $J_2 = (I, g)$. Then since $f \notin I(X)$, it follows that $V(J_1) \subsetneqq X$, and similarly $V(J_2) \subsetneqq X$, and so $X = V(J_1) \cup V(J_2)$ is reducible. The converse is similar (see Ex. 5.1).   Q.E.D.

5.8 The Nullstellensatz and Spec *A*

Corollary *Let k be an algebraically closed field. Then V and I induce one-to-one correspondences*

$$
\left\{ \begin{array}{c} \text{radical ideals } J \text{ of} \\ k[X_1, \ldots, X_n] \end{array} \right\} \quad \leftrightarrow \quad \left\{ \text{varieties } X \subset k^n \right\}
$$

$$
\cup \qquad\qquad\qquad\qquad \cup
$$

$$
\left\{ \begin{array}{c} \text{prime ideals } P \text{ of} \\ k[X_1, \ldots, X_n] \end{array} \right\} \quad \leftrightarrow \quad \left\{ \begin{array}{c} \text{irreducible varieties} \\ X \subset k^n \end{array} \right\}
$$

Therefore,

$$
\operatorname{Spec} k[X_1, \ldots, X_n] = \left\{ \text{irreducible varieties } X \subset k^n \right\}.
$$

Proposition *Let $A = k[x_1, \ldots, x_n]$ be a finitely generated k-algebra (where k is an algebraically closed field); write J for the ideal of relations holding between $x_1, \ldots, x_n$, so that $A = k[X_1, \ldots, X_n]/J$. Then there is a one-to-one correspondence*

$$
\operatorname{Spec} A \leftrightarrow \left\{ \text{irreducible subvarieties } X \subset V(J) \right\}.
$$

Proof We already know by Proposition 5.3 that maximal ideals correspond one-to-one with points of $V(J)$. Prime ideals of A correspond to prime ideals of $k[X_1, \ldots, X_n]$ containing J; so by Corollary 5.8, every prime ideal P of A is of the form $P = I(X) \bmod J$ for an irreducible variety $X \subset k^n$ with $J \subset P = I(X)$. This condition is equivalent to $V(J) \supset V(P) = V(I(X)) = X$.   Q.E.D.

5.9 The Zariski topology on a variety

It is easy to see that varieties $X \subset k^n$ form the closed sets of a topology on k^n, since

(i) the union of two varieties is again a variety:

$$
V(I) \cup V(J) = V(I \cap J) = V(IJ);
$$

(ii) if $X_i = V(J_i)$ is any family of varieties then their intersection is again a variety: $\bigcap X_i = V(\sum J_i)$.

The *Zariski topology* on a variety Y is defined by taking the subvarieties $X \subset Y$ as the closed sets. This is quite different from the usual (Euclidean) topology of $\mathbb{R}^n$ or $\mathbb{C}^n$. Subvarieties of $\mathbb{R}^n$ or $\mathbb{C}^n$ are of course closed in the Euclidean topology, because polynomials are continuous functions. But the Zariski topology is much weaker: a small Euclidean ball is not the complement of an algebraic variety, so not a Zariski open set.

The Zariski topology is a very weak topology, that is, it has very few closed sets: for example, in the case $n = 1$, any nonzero ideal of the polynomial ring $k[X]$ is of the form (f), and $V(f) = \{\text{roots of } f\} \subset k$ is a finite set. This is the cofinite topology on k (the nonempty open sets are the complements of finite sets): it is obviously not Hausdorff. In fact the topology of an irreducible variety X is the opposite of Hausdorff: since X is not the union of two proper closed subsets, any two nonempty open sets have nonempty intersection.

5.10 The Zariski topology on a variety is Noetherian

Proposition

(i) *Any decreasing chain $V_1 \supset V_2 \supset \cdots$ of varieties of k^n eventually stops.*

(ii) *Any nonempty set Σ of varieties of k^n has a minimal element.*

A topological space X is *Noetherian* if its closed sets have the d.c.c. Thus the proposition states that the Zariski topology on k^n is Noetherian.

Proof As in 3.1, (i) $\iff$ (ii) is formal. To prove (i), note that a decreasing chain $V_1 \supset V_2 \supset \cdots$ of varieties of k^n corresponds to an increasing chain $I(V_1) \subset I(V_2) \subset \cdots$ of ideals of $k[X_1, \ldots, X_n]$. By the Noetherian property of $k[X_1, \ldots, X_n]$, this chain stops with $I(V_k) = I(V_{k+1}) = \cdots$, and therefore $V_k = V_{k+1} = \cdots$.   Q.E.D.

5.11 Decomposition into irreducibles

Theorem *Let $X \subset k^n$ be a variety; then X has a decomposition*

$$X = X_1 \cup X_2 \cup \cdots \cup X_k, \tag{*}$$

with each X_i irreducible. By omitting some of the terms if necessary, I can arrange that the expression (*) satisfies $X_i \not\subset X_j$ for $i \neq j$, and then it is unique up to renumbering the components.

The X_i are the *irreducible components* of the variety X. In the case of a hypersurface $X = V(f) \subset k^n$, where k is algebraically closed, the components of X correspond to the distinct irreducible factors of f (see Ex. 5.4). The irreducible decomposition of a variety X generalises the idea of factorising an element of a ring as a product of irreducible factors.

Proof If X is not irreducible then $X = X_1 \cup X_2$ with $X_1, X_2 \subsetneq X$. Continuing in the same way with X_1 and X_2 gives a strictly decreasing chain of subvarieties of X, which must stop by the Noetherian property of the topology.

Here is a neat way of doing the Noetherian induction: let Σ be the set of varieties with no decomposition of the form (∗). The assertion is that $\Sigma = \emptyset$. If $\Sigma \neq \emptyset$ then, by the Noetherian property, Σ has a minimal element, say X; now X is not itself irreducible, so $X = X_1 \cup X_2$ with $X_1, X_2 \subsetneq X$. However, by the minimality of X, each of X_1 and X_2 does have a decomposition (∗), and putting them together gives one for X. This contradiction proves that $\Sigma = \emptyset$. The uniqueness statement is easy.   Q.E.D.

Corollary *A radical ideal J of $k[X_1, \ldots, X_n]$ (see 1.12) is an intersection of finitely many prime ideals.*

Proof Indeed, if the variety $X = V(J)$ has the irreducible decomposition (∗) then obviously

$$I(X) = \bigcap_{i=1}^{n} I(X_i). \qquad (**)$$

Now each $I(X_i)$ is a prime ideal by Proposition 5.7, and the Nullstellensatz, together with the assumption that J is radical, gives that $J = I(X)$.   Q.E.D.

5.12 The Zariski topology on a general Spec *A*

Now I want to give some idea of the geometry of Spec *A* for any ring *A*. As a set,

$$\text{Spec } A = \{\text{prime ideals of } A\}.$$

Definition The *Zariski topology* of Spec A is the topology whose closed sets are of the form

$$\mathcal{V}(I) = \{P \in \operatorname{Spec} A \mid P \supset I\}$$

where I is an ideal of A.

Note that $I \subset P$ means that every $f \in I$ maps to 0 in the map $A \to A/P$; I have already introduced in 1.3 the formal point of view that points of Spec A are in one-to-one correspondence with homomorphisms of A to fields, with P corresponding to the composite $A \twoheadrightarrow A/P \hookrightarrow k(P) = \operatorname{Frac}(A/P)$. If I write $f(P)$ for the image of f and call it "evaluation at P", then $\mathcal{V}(I) = \{P \in \operatorname{Spec} A \mid f(P) = 0 \text{ for all } f \in I\}$, by analogy with the definition of a variety $V(I)$ in 5.3. I will sometimes refer to $\mathcal{V}(I)$ as the "variety" of I (to mean closed set).

Why does this define a topology? Obviously, $\bigcap_{\alpha \in A} \mathcal{V}(I_\alpha) = \mathcal{V}(\sum I_\alpha)$, and both $\emptyset$ and Spec A are of the form $\mathcal{V}(I)$. As in 5.9 above, it is not hard to see that

$$\mathcal{V}(I_1) \cup \mathcal{V}(I_2) = \mathcal{V}(I_1 I_2) = \mathcal{V}(I_1 \cap I_2).$$

Indeed, the inclusions $\subset$ are obvious; conversely, if P is a prime ideal and $P \notin \mathcal{V}(I_1) \cup \mathcal{V}(I_2)$ then P does not contain either I_1 or I_2, and I can choose $f \in I_1 \setminus P$ and $g \in I_2 \setminus P$ to get the product $fg \in I_1 \cap I_2$ and $fg \notin P$, hence $P \notin \mathcal{V}(I_1 \cap I_2)$.

A closed set can be written in the form $\mathcal{V}(I)$ in more than one way: notice that $\mathcal{V}(I) = \mathcal{V}(\operatorname{rad} I)$. You can pass back from subsets of Spec A to ideals of A by taking intersections of prime ideals, that is, sending $X \subset \operatorname{Spec} A$ to $\mathcal{I}(X) = \bigcap_{P \in X} P$. By Corollary 1.12, if $V = \mathcal{V}(I)$ then $\operatorname{rad} I = \bigcap_{P \in V} P$, so that V can be written uniquely as $\mathcal{V}(I)$ with I a radical ideal. Thus $\mathcal{V}$ is a one-to-one correspondence from radical ideals of A to closed sets of Spec A. Notice that this is a formal analogue of the Nullstellensatz which is not deep, depending only on the easy result Corollary 1.12.

We have the same definition of irreducible closed set, and by copying the proof of Proposition 5.7, we can prove that $X = \mathcal{V}(J)$ is irreducible if and only if $\operatorname{rad} J = \mathcal{I}(X) = \bigcap_{P \in X} P$ is prime (see Ex. 5.8).

5.13 Spec *A* for a Noetherian ring

If A is a Noetherian ring then the Zariski topology on Spec A is Noetherian, since its closed sets are of the form $\mathcal{V}(I)$ for suitable ideals I. Also, the same argument as in Theorem 5.11 proves that any closed subset of Spec A is the union of a finite number of irreducible closed sets.

Corollary

(i) *If I is an ideal of a Noetherian ring A then the set of prime ideals containing I has a finite number of minimal elements P_i.*

(ii) $\operatorname{rad} I = \bigcap P_i$, *where the intersection is taken over the finitely many P_i of (i).*

(iii) *In particular, a Noetherian ring A containing zerodivisors either has nonzero nilpotents or has a finite number ≥ 2 of minimal primes.* Compare the discussion in 1.11 and Ex. 1.13.

Proof (i) The prime ideals containing I are in one-to-one correspondence with irreducible subvarieties of $\mathcal{V}(I)$, and the irreducible components of $\mathcal{V}(I)$ are exactly the maximal elements among these.

(ii) By Corollary 1.12, $\operatorname{rad} I = \bigcap P$ with the intersection taken over all primes containing I; but by (i), I can just take the intersection over the finitely many minimal ones.

(iii) The proof is as in 1.11. If $\operatorname{nilrad} A \neq 0$ then A has nilpotents. The alternative is $0 = \operatorname{rad} 0 = \bigcap P_i$, with P_i the minimal prime ideals of A. If there is only 1 of these then $0 = P_1$ is prime, so A is an integral domain. Q.E.D.

Remark The information contained in the Zariski topology of Spec A for a Noetherian ring A is neither more nor less than the inclusion relations between prime ideals. More precisely, a subset $X \subset \operatorname{Spec} A$ is closed in the Zariski topology if and only if the following two conditions hold:

(a) $P \in X \implies \mathcal{V}(P) \subset X$. In other words, for $P, Q \in \operatorname{Spec} A$ with $P \subset Q$, if $P \in X$ then also $Q \in X$.

(b) There are only finitely many minimal elements $P_i \in X$.

In other words, a Zariski closed set $X = \bigcup \mathcal{V}(P_i)$ consists of a finite set of primes P_i, together with all the primes Q containing any of these. The proof is an exercise in the formalism, see Ex. 5.10.

5.14 Varieties versus Spec A

Definition Let k be an algebraically closed field; a *geometric ring* is a finitely generated k-algebra A which is reduced (no nonzero nilpotents). That is, $A = k[X_1, \ldots, X_n]/I$, where $I = \operatorname{rad} I$ is a radical ideal.

To conclude this section, I want to compare and contrast briefly the variety $X = V(I) \subset k^n$ associated with a geometric ring $k[X_1, \ldots, X_n]/I$, and the construction Spec A for a general ring A.

The geometric case For a geometric ring A, the formal set-up is as follows:

1. The maximal spectrum is a variety m-Spec $A = V = V(I) \subset k^n$ by Proposition 5.3; Spec A itself is determined in terms of V as the set of irreducible subvarieties of V.

2. A is a ring of functions on V. Indeed, $A = k[X_1, \ldots, X_n]/I$, where $I = \operatorname{rad} I = \{$polynomials vanishing at all $P \in V\}$. So any $f \in A$ can be viewed as a polynomial function $f \colon V \to k$; that is, a function that can be written as a polynomial in the coordinates x_i of the ambient space k^n. In this connection, A should be thought of as determined by V, and we write $A = k[V]$, and call it the *coordinate ring* of V.

3. V has a Zariski topology: for $J \subset A$, let $\widetilde{J} \subset k[X_1, \ldots, X_n]$ be the ideal corresponding to J; then $V(J) = V(\widetilde{J}) \subset V(I)$. Moreover, the topology is defined by ideals of A, as follows: if we identify a point $P \in V$ with the maximal ideal $m_P = (x_1 - a_1, \ldots, x_n - a_n)$ of A as in Proposition 5.3, then

$$P \in V(J) \iff m_P \supset J.$$

4. The Nullstellensatz 5.6 says that $\bigcap_{Q \in V} m_Q = 0$; moreover, for any $P \in \operatorname{Spec} A$,

$$\bigcap_{Q \in V(P)} m_Q = P.$$

5. Polynomial maps: if $V \subset k^n$ and $W \subset k^m$ are varieties then algebraic geometry tells us that a polynomial map $f \colon V \to W$ is determined by a unique k-algebra homomorphism $\Phi \colon k[W] \to k[V]$, and conversely. For details, see any textbook on algebraic geometry, for example, [UAG], Proposition 4.5.

The general case Now let A be an arbitrary ring, and examine the extent to which Spec A can provide analogues of the above statements.

I Spec A is a set; in general we don't know anything about the relation between Spec A and m-Spec A, so that we are obliged to work with Spec A.

II An analogue of (2): for any $P \in$ Spec A, define $f(P)$ to be the residue of f modulo P,

$$f: P \mapsto f(P) = f \bmod P \in A/P \subset \mathrm{Frac}(A/P) = k(P).$$

In general this is a rather peculiar "function", since for different P it takes values in different fields $k(P)$, a bit like a vector field on a manifold M, whose value at a point P is an element of the tangent space $T_P M$. So for example, $n \in \mathbb{Z}$ is the "function" on Spec $\mathbb{Z}$ given by

$$\begin{cases} (0) \mapsto n \in \mathbb{Q} \\ p \mapsto n \bmod p. \end{cases}$$

However, in the geometric case, this can be identified with the way in which $f \in k[V]$ is viewed as a function on V:

$$f(a_1, \dots, a_n) = f \bmod m_P, \quad \text{where } m_P = (x_1 - a_1, \dots, x_n - a_n).$$

The special feature of the geometric case, as mentioned in the proof of Corollary 5.2, is that the ring $A = k[V]$ contains a field k mapping isomorphically under each quotient map $k[V] \to k[V]/m$. Thus the residue fields $k(m)$ are all identified.

III The Zariski topology on Spec A is an exact analogue of the Zariski topology on V.

IV The weak analogue of the Nullstellensatz is Corollary 1.10 and Corollary 1.12:

$$\bigcap_{P \in \mathrm{Spec}\, A} P = \mathrm{rad}\, 0 = \mathrm{nilrad}\, A.$$

That is, in the functional language of (ii), if $f(P) = 0$ for all $P \in$ Spec A, then f is nilpotent. Or in other words, if A is reduced, an element $f \in A$ is uniquely determined by the corresponding function $P \mapsto f(P) \in k(P)$ for all $P \in$ Spec A.

V If $\rho: B \to A$ is a ring morphism then ${}^a\rho:$ Spec $A \to$ Spec B is defined by $P \mapsto \rho^{-1}(P)$. This is continuous in the Zariski topology, and is a close analogue of (5). See Ex. 5.11.

Remark A little imagination is called for in this analogy: you have to draw Spec $\mathbb{Z}$ as the points $2, 3, 5, 7, \ldots, 163, \ldots, (0)$ strung out as a "line", with the generic point (0) as an everywhere dense "laundry mark", and look back at the weird pictures of Figures 0.7 and 1.6. Actually, one essential ingredient is still missing from the analogy, namely, the local ring A_P and the localisation map $f \mapsto f/1 \in A_P$ corresponding to every prime $P \in$ Spec A. Find out all about this in the next thrilling chapter.

Exercises to Chapter 5

5.1   Complete the proof of Proposition 5.7: $I(X)$ is prime implies that $X \subset k^n$ is irreducible.

5.2   Describe the irreducible components of $V(J) \subset k^3$, for each of the following 3 ideals $J \subset k[X, Y, Z]$:

   (1) $(Y^2 - X^4, X^2 - 2X^3 - X^2Y + 2XY + Y^2 - Y)$,

   (2) $(XY + YZ + XZ, XYZ)$

   (3) $((X - Z)(X - Y)(X - 2Z), X^2 - Y^2Z)$.

5.3   For each of the varieties $V(J)$ of Ex. 5.2, find if possible an element $f \in I(V(J))$ with $f \notin J$.

5.4   Let k be algebraically closed. For $f \in k[X_1, \ldots, X_n]$, write $V(f) \subset k^n$ for the hypersurface defined by $f = 0$. Prove that if f is irreducible and $f \nmid g$ then $V(f) \not\subset V(g)$. [Hint: there are two different methods of proof: either use the NSS, or eliminate one of the variables as in 1.5.] Deduce that the irreducible components of a hypersurface $V(g) \subset k^n$ are of the form $V(f_i)$ where $g = \text{const} \cdot \prod f_i^{n_i}$.

5.5   Let k be algebraically closed and $F \in k[X_1, \ldots, X_n]$ a reduced polynomial (that is, $F = \prod f_i$ with $f_i \nmid f_j$ distinct irreducible factors). Set $V = V(F) \subset k^n$. Prove that F generates $I(V)$.

5.6   Let k be a field, not necessarily algebraically closed. For an ideal $J \subset k[X_1, \ldots, X_n]$, and an extension field $k \subset K$, define a K-*valued point* of $V(J)$ to be a point $(a_1, \ldots, a_n) \in K^n$ such that $f(a_1, \ldots, a_n) = 0$ for all $f \in J$. State and prove a version of the Nullstellensatz in terms of K-valued points of $V(J)$ for all algebraic extension fields K of k. (Compare 5.4.)

5.7   This exercise is for the student who knows Galois theory. Let k be a field and $k \subset K$ a Galois field extension with Galois group

$G = \text{Gal}(K/k)$. Prove that two K-valued points $(a_1, \ldots, a_n)$ and $(b_1, \ldots, b_n)$ of $V(J)$ correspond to the same maximal ideal of $k[X_1, \ldots, X_n]$ if and only if there is an element $\sigma \in G$ such that $(a_1, \ldots, a_n) = (\sigma(b_1), \ldots, \sigma(b_n))$. [Hint: how do you do this if $n = 1$?]

5.8   If A is a Noetherian ring, prove that a closed set $X \subset \text{Spec } A$ is irreducible if and only if $\mathcal{I}(X) = \bigcap_{P \in X} P$ is prime. [Hint: set $I = \mathcal{I}(X)$. As in the proof of Proposition 5.7, suppose I is not prime, so there exists $f, g \notin I$ with $fg \in I$. Consider

$$X = \mathcal{V}(I, f) \cup \mathcal{V}(I, g).$$

Conversely, if $X = X_1 \cup X_2$ with strictly smaller closed sets X_1, X_2, you have to show that $I = \mathcal{I}(X_1) \cap \mathcal{I}(X_2)$ is not prime.]

5.9   Prove that $\mathcal{V}(P) \subset \text{Spec } A$ is the closure in the Zariski topology of the point P.

5.10   Prove Remark 5.13.

5.11   If $\varphi \colon B \to A$ is a ring homomorphism, show that ${}^a\varphi \colon \text{Spec } A \to \text{Spec } B$ defined by ${}^a\varphi(P) = \varphi^{-1}(P)$ is well defined and continuous for the Zariski topology.

5.12   Let A and B be geometric rings over an algebraically closed field k, and $\varphi, {}^a\varphi$ as in Ex. 5.11. Describe ${}^a\varphi$ as a polynomial map between the varieties V and W corresponding to A and B.

6

Rings of fractions $S^{-1}A$
and localisation

The material of this chapter is rather elementary, and most of it could have been included more-or-less anywhere earlier in the course. As mentioned in 1.13, localisation is a notion that allows us to reduce many questions of commutative algebra to the case of local rings. The ideas and results of this section, particularly localisation at a prime, are used throughout Chapters 7 and 8.

6.1 The construction of $S^{-1}A$

Given an integral domain A and a subset S not containing 0, we can consider the subring

$$A[S^{-1}] = A[\{1/s | s \in S\}] \subset K,$$

where $K = \operatorname{Frac} A$ is the field of fractions of A. The construction of $S^{-1}A$ is slightly more general, because it allows S to have zerodivisors.

Definition Let A be a ring and $S \subset A$ a multiplicative set (that is, $1 \in S$, and $st \in S$ for all $s, t \in S$). Introduce the following relation $\sim$ on $A \times S$:

$$(a, s) \sim (b, t) \iff \exists u \in S \text{ such that } u(at - bs) = 0;$$

it will be proved shortly that $\sim$ is an equivalence relation. Write a/s for the class of (a, s). Note that the more simple-minded relation $at = bs$ is not an equivalence relation, see Example 1 below and Ex. 6.1. Then the *ring of fractions* of A with respect to S is

$$S^{-1}A = (A \times S)/\sim$$

with ring operations defined by the usual arithmetic operations on fractions:

$$\frac{a}{s} \pm \frac{b}{t} = \frac{(at \pm bs)}{st}, \qquad \text{and} \qquad \frac{a}{s} \cdot \frac{b}{t} = \frac{ab}{st}.$$

Proposition

    (i) $\sim$ *is an equivalence relation.*
    (ii) *The ring operations are well defined, and $S^{-1}A$ is a ring.*
    (iii) $\varphi \colon A \to S^{-1}A$ *given by $a \mapsto a/1$ is a ring homomorphism.*

The two most popular and useful choices of multiplicative sets are $S = \{1, f, f^2, \ldots\}$ for an element $f \in A$ (considered in more detail in Lemma 6.2) and $S = A \setminus P$ for a prime ideal P (see 6.4).

Proof This is all just tedious verification. Suppose that $(a, s) \sim (a', s') \sim (a'', s'')$. Then

$$\exists u \text{ such that } u(as' - a's) = 0 \quad \text{and } \exists u' \text{ such that } u'(a's'' - a''s') = 0,$$

and therefore

$$uu's'(as'' - a''s) = u's''(u(as' - a's)) + us(u'(a's'' - a''s')) = 0.$$

Therefore $(a, s) \sim (a'', s'')$, because $uu's' \in S$.

    Similarly, it is not hard to see that the arithmetic operations are well defined and define a ring structure on $S^{-1}A$. For example,

$$(a, s) \sim (a', s') \implies \exists u \in S \text{ such that } u(as' - a's) = 0.$$

Therefore for any (b, t)

$$0 = ubt(as' - a's) = u(abs't - a'bst) \implies ab/st = a'b/s't.$$

(iii) is similarly easy.   Q.E.D.

Example 1 Consider $A = k[X, Y]/(XY)$ and $S = \{1, X, X^2, \ldots\}$. Then $S^{-1}A = k[X, X^{-1}]$; note that

$$\varphi(A) = k[X] \subset k[X, X^{-1}].$$

So φ kills off Y first, then just add in X^{-1} to the integral domain $k[X]$.

Example 2 Consider $A = k[X,Y]/(Y^2)$; then $k(X)[Y]/(Y^2)$ is the ring of fractions $S^{-1}A$, where $S = \{a(X) + b(X)Y \mid a(X) \neq 0\}$.

6.2 Easy properties

(a) $\ker \varphi = \{a \in A \mid \exists s \in S \text{ such that } sa = 0\}$; this is clear from the definition of $\sim$.

(b) $S^{-1}A = 0 \iff 0 \in S \iff S$ contains a nilpotent element.

(c) If $S^{-1}A \neq 0$ then $\varphi(S)$ consists of units (since $(1/s) \cdot (s/1) = 1$). In fact $\varphi \colon A \to S^{-1}A$ is the universal ring with this property. The point is just that if $f \colon A \to B$ is a ring homomorphism and $f(S)$ consists of units, then there exists a unique ring homomorphism $f' \colon S^{-1}A \to B$ with the property that f is the composite $A \to S^{-1}A \to B$ (of course, f' is given by $a/s \mapsto f(a)/f(s)$). The ring $S^{-1}A$ and the homomorphism are uniquely determined by this property (see Ex. 6.11).

Lemma *For $f \in A$, write $S = \{1, f, f^2, \dots\}$, and $A_f = S^{-1}A$. Then*

$$A_f \cong A[X]/(Xf - 1).$$

Proof Define a ring homomorphism $\alpha \colon A[X] \to A_f$ by $\alpha(a) = \varphi(a) = a/1$ for $a \in A$ and $\alpha(X) = 1/f$; then α is clearly surjective. I have to prove that $\ker \alpha = (Xf - 1)$. The inclusion $\supset$ is obvious. So take $h \in \ker \alpha$, and let's prove that $h \in (Xf - 1)$.

Step 1 I claim first that $f^n \cdot h(X) \in (Xf - 1)$ for some n.

To see this in more detail, note that $h = h(X)$ satisfies $h(1/f) = 0 \in A_f$, that is, $f^n \cdot h(1/f) = 0 \in A$ for some $n \geq \deg f$. Then $f^n \cdot h(X) = G(fX)$ where $G = G(Y) \in A[Y]$ satisfies $G(1) = 0$. By the remainder theorem, $G(Y) = (Y - 1) \cdot G_1(Y)$ for some polynomial $G_1(Y)$, so that $f^n \cdot h(X) = G(fX) = (fX - 1) \cdot G_1(fX)$.

Step 2 The point now is that f and $Xf - 1$ are coprime, so that

$$f^n \cdot h(X) \in (Xf - 1) \implies h(X) \in (Xf - 1).$$

Indeed, $1 = Xf - (Xf - 1)$, so that by taking nth powers and using the binomial theorem, I get

$$1 = X^n f^n + a \cdot (Xf - 1) \qquad \text{for some } a \in A[X].$$

Therefore $h(X) = X^n f^n \cdot h(X) + a \cdot (Xf - 1) \cdot h(X) \in (Xf - 1)$.   Q.E.D.

6.3 Ideals in A and $S^{-1}A$

Quite generally, given a ring homomorphism $\varphi \colon A \to B$, there is a correspondence

$$e \colon \{\text{ideals of } A\} \to \{\text{ideals of } B\} \quad \text{given by } e(I) = \varphi(I)B = IB$$

(called "extension") and

$$r \colon \{\text{ideals of } B\} \to \{\text{ideals of } A\} \quad \text{given by } r(J) = \varphi^{-1}J$$

(called "restriction", and often written $A \cap J$). Now set $B = S^{-1}A$. I sometimes write $S^{-1}I = e(I) = \varphi(I) \cdot B$.

Proposition

   (a) *For any ideal J of $S^{-1}A$, I have $e(r(J)) = J$.*
   (b) *For any ideal I of A, I have*

$$r(e(I)) = \{a \in A \mid as \in I \text{ for some } s \in S\}.$$

   (c) *If P is prime and $P \cap S = \emptyset$ then $e(P) = S^{-1}P$ is a prime ideal of $S^{-1}A$.*

Proof (a) If $b/s \in J$ then $b \in \varphi^{-1}J$, so $b/s \in e(r(J))$; this proves $J \subset e(r(J))$, and the inclusion the other way round is trivial.
   (b) If $a \in r(e(I))$ then $a/1 = b/t \in S^{-1}A$ for some $b \in I$, $t \in S$. Now there exists $u \in S$ such that $uta = ub \in I$, so that $s = ut \in S$ knocks a into I. The inclusion the other way round is again trivial.
   (c) is an easy exercise based on (b).   Q.E.D.

Corollary

   (i) *For an ideal I of A, the necessary and sufficient condition for $r(e(I)) = I$ is*

$$as \in I \implies a \in I \qquad \text{for all } s \in S. \qquad (*)$$

   (ii) *Therefore e and r define inverse one-to-one correspondences*

$$\{\text{ideals of } A \text{ satisfying } (*)\} \leftrightarrow \{\text{ideals of } S^{-1}A\}.$$

   It follows in particular that $S^{-1}A$ is Noetherian if A is.
   (iii) $r(e(I)) = A \iff e(I) = S^{-1}A \iff I \cap S \neq \emptyset$.

(iv) *If P is a prime ideal of A such that $P \cap S = \emptyset$ then (*) holds for P, and $r(e(P)) = P$; therefore r: Spec $S^{-1}A \hookrightarrow$ Spec A identifies Spec $S^{-1}A$ with the subset $\{P \in$ Spec $A \mid P \cap S = \emptyset\} \subset$ Spec A.*

This all follows directly from the proposition. A slightly different description of $r(e(I))$ in terms of primary decomposition is given later (see Proposition 7.13).

6.4 Localisation

Localisation is a particular case of ring of fractions; a slightly less general case of it was discussed in 1.13. If P is a prime ideal then $S = A \setminus P$ is a multiplicative set; set $A_P = S^{-1}A$.

Proposition *$a/s \in A_P$ is a unit of $A_P \iff a \notin P$. Therefore A_P is a local ring, with maximal ideal $e(P) = PA_P$.*

The local ring (A_P, PA_P) is called the *localisation* of A at P.

Proof $\impliedby$ is obvious since $(a/s) \cdot (s/a) = 1$.
   $\implies$ If $(a/s) \cdot (b/t) = 1$, there exists $u \in S$ such that $u(st - ab) = 0$, which implies that $uab = ust \notin P$. Therefore, because P is an ideal, also $a \notin P$. Q.E.D.

Recall from 1.13 and Ex. 1.10 that a local ring (A, m) is characterised in any of the three following ways:

(a) A has a unique maximal ideal m;
(b) $m = \{$nonunits of $A\}$ is an ideal;
(c) $m \subset A$ is a maximal ideal and $x \in m \implies 1 + x$ is a unit.

Examples Two standard examples were given in 1.14:

$$\mathbb{Z}_{(p)} = \{a/b \in \mathbb{Q} \mid a, b \in \mathbb{Z} \text{ such that } p \nmid b\}$$

for p a prime number, and

$$k[X]_{(X-a)} = \{f/g \in k(X) \mid (X - a) \nmid g\}$$

for $a \in k$. The second of these can be described as the subring of the rational function field $k(X)$ consisting of functions which are regular (that is, well defined) at the point $a \in k$. Thus it is a ring of germs of functions defined locally near $a \in k$. Other local rings of germs of functions were described in 1.15 and 3.3, (iii).

More generally, if k is algebraically closed, $P \in \operatorname{Spec} k[X_1, \ldots, X_n]$ a prime ideal and $V \subset k^n$ the corresponding irreducible variety, the localisation $k[X_1, \ldots, X_n]_P$ of the polynomial ring at P is the subring of the rational function field $k(X_1, \ldots, X_n)$ consisting of quotients $h = f/g$ with $g \not\equiv 0$ on V. Thus h is regular at "most" points of V; in fact, the set of points at which it is regular is a dense open set in the Zariski topology of V. So "local ring" means "ring of rational functions defined locally near a general point of V".

6.5 Modules of fractions

Proposition *Let A be a ring, S a multiplicative set of A, and $S^{-1}A$ the ring of fractions of A with respect to S. Then modules over $S^{-1}A$ can be naturally identified with A-modules M having the property that the multiplication map $\mu_s \colon M \to M$ is bijective for all $s \in S$.*

Proof If M is an $S^{-1}A$-module then I can consider it as an A-module by $(a, m) \mapsto (a/1) \cdot m$. Then multiplication by $s \in S$ is obviously bijective. You can see the converse by looking at the structure map $A \to \operatorname{End} M$ (as in 2.2) and muttering something about the universal property of $A \to S^{-1}A$ as in 6.2, (c).

More concretely, if M is an A-module such that $\mu_s \colon M \to M$ is bijective for all $s \in S$, then I can define an action of $S^{-1}A$ on M by setting $(a/s) \cdot m = a \cdot \mu_s^{-1}(m)$. This is well defined: $a/s = b/t$ implies that there exists $u \in S$ such that $uat = ubs$, and therefore $ub\mu_s = ua\mu_t$ as maps $M \to M$, and $b\mu_s = a\mu_t$, because multiplication by u is one-to-one. Q.E.D.

Definition Let M be an A-module, $S \subset A$ a multiplicative set; then $S^{-1}M$ is the $S^{-1}A$-module defined as follows: define the equivalence relation $\sim$ on $M \times S$ as before

$$(m, s) \sim (n, t) \iff \exists u \in S \text{ such that } utm = usn,$$

and set $S^{-1}M = (M \times S)/\sim$. Then define the module operations by $m/s \pm n/t = (mt \pm ns)/st$ and $(a/s) \cdot (n/t) = an/st$. Checking that this works is just a repetition of what was said (and left unsaid) in 6.1.

If $S = A \setminus P$ then $S^{-1}M$ is a module over the local ring $S^{-1}A = A_P$, and is also written $S^{-1}M = M_P$.

Remarks

(i) $M \to S^{-1}M$ has the universal property (as in 6.2, (c)) for homomorphisms of M to an $S^{-1}A$-module.

(ii) If you know about tensor products, then $S^{-1}M = S^{-1}A \otimes_A M$; if you don't, ignore this remark.

6.6 Exactness of S^{-1}

The construction of $S^{-1}M$ is functorial in M: in other words, an A-linear map $g: M \to N$ induces a map $g': S^{-1}M \to S^{-1}N$ by the formula $g'(m/s) = g(m)/s$.

Proposition *If* $L \xrightarrow{\alpha} M \xrightarrow{\beta} N$ *is an exact sequence then so is*

$$S^{-1}L \xrightarrow{\alpha'} S^{-1}M \xrightarrow{\beta'} S^{-1}N.$$

This property is expressed by saying that S^{-1} is an *exact functor*. It follows in particular that localisation at a prime P preserves short exact sequences and direct sum of modules.

Proof Let $m/s \in S^{-1}M$. Then

$$\beta'(m/s) = 0 \iff \exists u \in S \text{ such that } u\beta(m) = 0$$
$$\iff \exists u \in S \text{ such that } \beta(um) = 0.$$

Using the fact that $\operatorname{im}\alpha = \ker\beta$ in the original sequence gives

$$\iff \exists u \in S \text{ and } \exists n \in L \text{ such that } um = \alpha(n)$$
$$\iff \exists u \in S \text{ and } \exists n \in L \text{ such that } m/s = \alpha'(n/us). \quad \text{Q.E.D.}$$

Corollary *(i) If* $L \subset M$ *and* $N = M/L$ *then*

$$S^{-1}L \subset S^{-1}M \qquad and \qquad S^{-1}N = S^{-1}M/S^{-1}L.$$

(ii) If $L, L' \subset M$ *are submodules then*

$$S^{-1}(L \cap L') = S^{-1}L \cap S^{-1}L' \subset S^{-1}M.$$

Proof (i) follows directly, (ii) by considering the exact sequence

$$0 \to L \cap L' \to L \to M/L'. \quad \text{Q.E.D.}$$

6.7 Localisation commutes with taking quotients

As usual, A is a ring, S a multiplicative set and I an ideal; write T for the image of S in A/I. Then $S^{-1}I = I \cdot S^{-1}A$ is an ideal of $S^{-1}A$, and I can take the quotient ring $S^{-1}A/S^{-1}I$. On the other hand, I can take the quotient A/I first, then localise to get $T^{-1}(A/I)$.

Corollary $T^{-1}(A/I) \cong S^{-1}A/S^{-1}I$. *In particular, for a prime ideal* P, *the field* A_P/PA_P *is equal to the field of fractions* $k(P) = \mathrm{Frac}(A/P)$ *of the integral domain* A/P.

In other words, we can construct the local ring A_P, PA_P as in 6.4 and then pass to its residue field; this gives the same thing as the residue field $k(P)$ of A at P discussed in 1.3 and 5.12.

Proof The quotient ring A/I can be viewed just as an A-module, and then the ring of fractions $T^{-1}(A/I)$ equals the module of fractions $S^{-1}(A/I)$. Corollary 6.6, (i) then gives an isomorphism of modules $T^{-1}(A/I) = S^{-1}(A/I) \cong S^{-1}A/S^{-1}I$, and it's easy to see that this is a ring homomorphism. Q.E.D.

You have already seen a particular case in 1.14: $\mathbb{Z}_{(p)}$ is a local ring with maximal ideal $p\mathbb{Z}_{(p)}$, and residue field $\mathbb{Z}_{(p)}/p\mathbb{Z}_{(p)} \cong \mathbb{Z}/(p) = \mathbb{F}_p$. Another case is when $A = k[X_1, \ldots, X_n]$ and $P = I(V)$ is the ideal defining

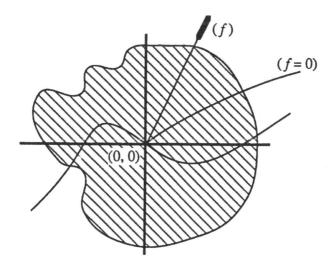

Figure 6.8. The spectrum of a local ring $\mathrm{Spec}\, A_m$

an irreducible variety V. Then the corollary says that the following two different procedures give the same field $k(V)$. First pass to the quotient ring $k[V] = A/I(V)$, which is the coordinate ring of V, the ring of polynomial functions on V. This is an integral domain since $I(V)$ is prime (see Proposition 5.7), so I can take its field of fractions $k(V) = \operatorname{Frac} k[V]$. Alternatively, take the localisation A_P, and then pass to the quotient ring A_P/PA_P. Here, as described at the end of 6.4, A_P is the subring of the rational function field $k(X_1, \ldots, X_n)$ consisting of fractions f/g with $g \notin I(V)$, and passing to the quotient ring means working modulo those fractions f/g with $f \in I(V)$, or in other words, only taking account of f/g as a function on V (more precisely, on a dense open set of V, since f/g is regular at $P \in V$ with $g(P) \neq 0$).

This field $k(V)$ can be viewed as the field of rational functions on V, and is called the *function field* of V.

6.8 Localise and localise again

If $P \subset Q$ are prime ideals then, of course, $P \cap (A \setminus Q) = \emptyset$, so that PA_Q is a prime ideal of A_Q by Proposition 6.3, (c). One sees easily that A_P is the localisation of A_Q at PA_Q. See Ex. 6.13 for details.

For example, let $A = k[X, Y]$ be the ring of polynomial functions on the (X, Y)-plane k^2, and $m = (X, Y)$ the maximal ideal of the origin $(0, 0)$. Then A_m is the local ring of $(0, 0) \in k^2$, consisting of all rational functions f/g with $g(0, 0) \neq 0$. Now although A_m has only one maximal ideal, $\operatorname{Spec} A_m$ is still an object with lots of substance (see Figure 6.8). Namely, A_m has a prime ideal for every irreducible curve through $(0, 0)$. Indeed, if $p \in m$ is an irreducible polynomial defining a curve $C : (p = 0)$ passing through $(0, 0)$, then $(p) \subset m$ and pA_m is a prime ideal. The localisation $A_{(p)} = (A_m)_{pA_m} = \{f/g \mid p \nmid g\} \subset k(X, Y)$ is a bigger ring than A_m, consisting of all rational functions f/g on k^2 defined at all sufficiently general points of C.

Exercises to Chapter 6

6.1   Find a ring A and a multiplicative set S such that the relation $(a, s) \sim (b, t) \iff at = bs$ is not an equivalence relation. [Hint: S must contain zerodivisors.]

6.2   Prove that addition $a/s + b/t = (at + bs)/st$ on $S^{-1}A$ is well defined, and that $S^{-1}A$ satisfies the distributive law.

6.3   (a) Let $A = A' \times A''$; prove that A' and A'' are rings of fractions of A.

         (b) Now suppose that A' and A'' are integral domains, and $A \subset A' \times A''$ a subring that maps surjectively to each factor; let S be a multiplicative set of A. State and prove the necessary and sufficient condition that S must satisfy in order for $S^{-1}A$ to be a ring of fractions of A'. [Hint: start by considering $k[X, Y]/(XY) \subset k[X] \times k[Y]$ as in Ex. 1.3.]

6.4   (a) Give an example of a ring A and distinct multiplicative sets S, T such that $S^{-1}A = T^{-1}A$.

         (b) Prove that for fixed S, there is a maximal multiplicative set T with this property, defined by

$$T = \{t \in A \mid at \in S \text{ for some } a \in A\}.$$

6.5   Find all intermediate rings $\mathbb{Z} \subset A \subset \mathbb{Q}$. [Hint: as a starter, consider the subring $\mathbb{Z}[2/3] \subset \mathbb{Q}$. Is $1/3 \in \mathbb{Z}[2/3]$?]

6.6   Let $M_i \subset M$ be submodules indexed by some set J, for which $M = \sum_{i \in J} M_i$. Suppose that S is a multiplicative set, and $S^{-1}M_i = 0$ for all $i \in J$. Make an original discovery concerning $S^{-1}M$.

6.7   Complete the proof from Ex. 4.12 that if $A \subset B$ is finite and P a prime ideal of A then there is at least one and at most finitely many prime ideals Q of B with $Q \cap A = P$. [Hint: you can reduce to the case $P = 0$ by passing to the quotient modulo P; now localise to reduce to the case that A is a field, and use Ex. 4.12, (ii).]

6.8   If $S = \{1, f, f^2, \dots\}$ is a multiplicative set of A, prove that $\operatorname{Spec}(A_f) \subset \operatorname{Spec} A$ is the complement of the closed set $\mathcal{V}(f)$ (see Corollary 6.3).

6.9 $A = k[V]$ is the coordinate ring of a variety $V \subset k^n$, and $f \in A$. Prove that $A[1/f]$ is the coordinate ring of a variety $V_f \subset k^{n+1}$, which is in natural one-to-one correspondence with the open set $V \setminus V(f)$. Compare [UAG], 4.13.

6.10   Prove that an element $f/g \in k[X_1, \dots, X_n]_P$ is a well-defined function on a Zariski open set of k^n that intersects $V = V(P) \subset k^n$ in a dense open set of V (see 6.4).

The remaining exercises of this section are concerned with the idea of universal mapping property (UMP) and its many elementary applications in algebra.

6.11   Prove that the ring $S^{-1}A$ and the homomorphism $A \to S^{-1}A$ is uniquely determined (up to isomorphism) by its UMP discussed in Remark 6.2, (c). [Hint: suppose that both $A \to A'$ and $A \to A''$ have the stated UMP for S. Prove that there exist mutually inverse maps $A' \to A''$ and $A'' \to A'$.]

6.12   Give an alternative proof of Lemma 6.2 based on the UMP of $S^{-1}A$.

6.13   Let $S \subset T$ be multiplicative sets of a ring A. Write $A' = S^{-1}A$ and $\varphi\colon A \to A'$ for the ring of fractions w.r.t. S and $T' = \varphi(T)$. Prove that $T^{-1}A = T'^{-1}A'$. In other words, a composite of two localisations is a localisation. [Hint: the easiest way to do this is to use the UMP of 6.2, (c) and Ex. 6.11.]

    As a particular case, prove that if $P \subset Q$ are prime ideals, then A_P is a localisation of A_Q at the prime ideal PA_Q. Compare 6.8.

6.14   The polynomial ring $A[X]$ has a homomorphism $A \to A[X]$ and a distinguished element $X \in A[X]$. Prove that $A \to A[X]$ has the UMP for a ring homomorphism $A \to B$ and an element $b \in B$, and is uniquely determined by it (up to isomorphism). [Hint: compare Ex. 0.4.] In other words, you can take this UMP as the definition of the polynomial ring $A[X]$, and regard all that mumbo-jumbo about sequences $a_0, a_1, \ldots, a_n \in A$ and formal ring operations etc. as a construction proving that the UMP has a solution.

6.15   Find a UMP description of the quotient ring A/I by an ideal I; show that you can take this as the definition of A/I, and regard all that mumbo-jumbo about equivalence relations and cosets etc. as a construction proving that the UMP has a solution. Give a proof of Proposition 6.7 based on putting together the UMPs of $S^{-1}A$ and A/I.

7

Primary decomposition

Factorisation into primes can be expressed in terms of intersection of ideals: if $n = \prod p_i^{n_i}$ is a prime factorisation in a UFD such as $\mathbb{Z}$ then $(n) = \bigcap(p_i^{n_i})$. Writing a variety $V = \bigcup V_i$ as a union of irreducible varieties expresses the radical ideal $I(V) \subset k[X_1, \ldots, X_n]$ as an intersection of primes $I(V) = \bigcap P_i$, where the $P_i = I(V_i)$ are the finitely many minimal primes containing I (see Theorem 5.11). Corollary 5.13, (ii) was a formal statement to the same effect for a radical ideal I of any Noetherian ring. This expression of an ideal as an intersection of prime ideals is an analogue of factorisation into primes that has extremely wide generalisations. To allow for ideals that are not radical, I have to generalise prime ideals to primary ideals, for example, an ideal such as (p^n) in $\mathbb{Z}$.

Another interpretation of factorisation into irreducibles, say in a UFD A, is to think of the factors of $f = \prod f_i^{n_i}$ as the elements of A that are zerodivisors for the module $M = A/(f)$, and the irreducible factors as the "extremal" zerodivisors, killing the smallest submodule of M, or generating the biggest ideal of A. This chapter starts with primary decomposition of modules in the style of Bourbaki, Supp and Ass and all that Jazz. We will see that if A is Noetherian, a finite A-module M "lives over" a closed subset $\operatorname{Supp} M$ of $\operatorname{Spec} A$, as depicted in the Frontispiece; we relate this to zerodivisors of M in A, and interpret the irreducible components of $\operatorname{Supp} M$ in terms of associated primes.

From 7.7 on, I return to the pre-Bourbaki tradition: primary decomposition in polynomial rings goes back to the 19th century German school, and the abstract algebraic treatment in 7.11, which is simpler and much more general, was Emmy Noether's first major contribution to algebra around 1920, and probably the main reason Noetherian rings are named after her.

7.1 The support of a module Supp M

Definition Let M be an A-module. The *support* of M is the subset

$$\text{Supp } M = \{P \in \text{Spec } A \mid M_P \ne 0\} \subset \text{Spec } A.$$

Recall that $M_P = S^{-1}M$ is the module of fractions with respect to $S = A \setminus P$.

For $m \in M$, the *annihilator* of m is the ideal $\text{Ann } m$ or $\text{Ann}_A m$ defined by $\text{Ann } m = \{f \in A \mid fm = 0\}$; similarly, $\text{Ann } M = \{f \in A \mid fM = 0\}$. An element $f \in A$ is a *zerodivisor* for M if $fm = 0$ for some $0 \ne m \in M$.

Recall from 5.12 that an ideal I of A corresponds to the subset $\mathcal{V}(I) = \{P \in \text{Spec } A \mid P \supset I\}$, which is by definition a closed set of the Zariski topology on $\text{Spec } A$.

Proposition

(i) *If M is a module generated by a single element x and $I = \text{Ann } x$ (so that $M \cong A/I$) then* $\text{Supp } M = \mathcal{V}(I)$.

(ii) $M = \sum_{i \in J} M_i \implies \text{Supp } M = \bigcup_{i \in J} \text{Supp } M_i$.

(iii) *If $L \subset M$ and $N = M/L$ then* $\text{Supp } M = \text{Supp } L \cup \text{Supp } N$.

(iv) *If M is finite over A then $\text{Supp } M = \mathcal{V}(\text{Ann } M)$, and is a closed subset of* $\text{Spec } A$.

(v) *If $P \in \text{Supp } M$ then* $\mathcal{V}(P) \subset \text{Supp } M$.

Proof (i) By definition of a module of fractions (see 6.5),

$$\frac{x}{1} = 0 \in M_P \iff sx = 0 \text{ for some } s \notin P$$
$$\iff (A \setminus P) \cap \text{Ann } x \ne \emptyset.$$

The really hard part of the proof is figuring out how to negate this complicated sentiment:

$$\frac{x}{1} \ne 0 \iff P \supset \text{Ann } x = I \iff P \in \mathcal{V}(I).$$

(ii) is clear: $M_i \subset M$, so that by Corollary 6.6, (i), $(M_i)_P \subset M_P$. Also if $(M_i)_P = 0$ for all $i \in I$ then every element of M is killed by some $s \notin P$, so that $M_P = 0$.

(iii) follows from Proposition 6.6, since $0 \to L \to M \to N \to 0$ exact implies that $0 \to L_P \to M_P \to N_P \to 0$ is exact for every $P \in \text{Spec } A$.

(iv) follows from (i) and (ii): if $\{m_1, \ldots, m_k\}$ is a family of generators of M then

$$\operatorname{Supp} M = \bigcup \operatorname{Supp} A m_i = \bigcup_i \mathcal{V}(\operatorname{Ann}(m_i))$$

$$= \mathcal{V}(\bigcap_i \operatorname{Ann}(m_i)) = \mathcal{V}(\operatorname{Ann} M),$$

where I'm using the obvious fact (compare Ex. 1.(v)(a) or 5.12) that for P a prime ideal, $P \supset \bigcap I_i$ if and only if $P \supset I_i$ for some i.

(v) is easy: $Q \in \mathcal{V}(P) \iff Q \supset P$, so that $A \setminus Q \subset A \setminus P$, hence M_P is a localisation of M_Q:

$$M_P = (A \setminus P)^{-1} M = (A \setminus P)^{-1}\big((A \setminus Q)^{-1} M\big) = (M_Q)_P.$$

(Compare 6.8.) Thus $M_P \neq 0$ implies that $M_Q \neq 0$.   Q.E.D.

7.2 Discussion

The idea of the above definition is as follows: corresponding to the A-module M, consider all the localisations M_P, which are modules over A_P. I picture the disjoint union $\mathcal{M} = \bigsqcup M_P$ as a family living over Spec A, as in Figure 7.2. The A_P-module M_P is called the *stalk* of $\mathcal{M}$ over P. This picture suggests at once a number of fun activities: for example, an element $m \in M$ gives rise to a family $\{m_P = m/1 \in M_P\}$ of elements, one in each stalk (see also Ex. 7.1).

The definition of Supp M is concerned simply with marking out the subset of Spec A where $M_P \neq 0$. Proposition 7.1, (iii) says that this is a closed subset of the Zariski topology (at least for a finite module M). As we know from Remark 5.13, a Zariski closed set $X \subset \operatorname{Spec} A$ is made up (at least in the Noetherian case) of a finite number of minimal primes P_i corresponding to the components $V_i = \mathcal{V}(P_i)$ of X, together with the points of the Zariski closure $\mathcal{V}(P_i)$ of P_i, that is, prime ideals $Q \supset P_i$.

We're going to see that, under finiteness assumptions on A and M, it's possible to associate in a natural way a finite number of primes P_i with a module M (determined in terms of the zerodivisors of M), such that

$$\operatorname{Supp} M = \bigcup \mathcal{V}(P_i).$$

For example, if A is an integral domain and $M = A$, then 0 is the only element of A that kills anything in M, and $(0) \in \operatorname{Supp} M$, since the localisation at (0) is just the field of fractions $A_{(0)} = \operatorname{Frac} A$. Then,

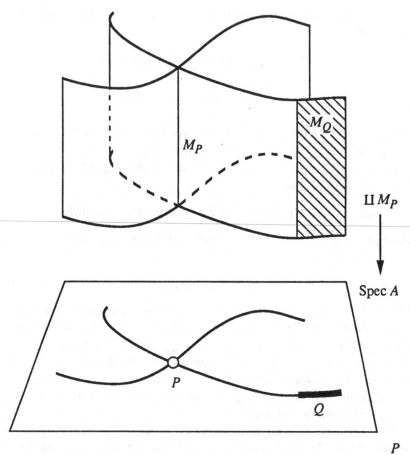

Figure 7.2. An A-module M gives rise to the family $\mathcal{M} = \{M_P\}$ for $P \in \operatorname{Spec} A$, which we can view as fibred over $\operatorname{Spec} A$.

of course, every $A_P \neq 0$, so that $\operatorname{Supp} A$ is the whole of $\operatorname{Spec} A$. The point of view adopted here is that the minimal prime (0) is the really important one, and all the other are in $\operatorname{Supp} M$ merely because they contain (0).

7.3 Definition of $\operatorname{Ass} M$

Definition Let A be a ring and M an A-module. An *assassin*† (or *associated prime*) of M is an ideal P of A such that

† An example of N. Bourbaki's sense of humour at its most exuberant.

(i) $P \in \operatorname{Spec} A$;

(ii) M contains a submodule isomorphic to A/P; or equivalently,

$$\exists x \in M \text{ such that } P = \operatorname{Ann} x.$$

Write $\operatorname{Ass} M = \{\text{assassins of } M\}$.

Each $P \in \operatorname{Ass} M$ obviously contains $\operatorname{Ann} M = \bigcap_{m \in M} \operatorname{Ann} m$.

Example Let $n \in \mathbb{Z}$, $n = p^\alpha q^\beta$ for two primes $p \neq q$; then the $\mathbb{Z}$-module $\mathbb{Z}/(n)$ has $\operatorname{Ass}(\mathbb{Z}/(n)) = \{(p), (q)\}$. For $m = p^{\alpha-1} q^\beta \bmod n \in \mathbb{Z}/(n)$ has $\operatorname{Ann} m = (p)$, and similarly for q; also, $\mathbb{Z}/(l) \subset \mathbb{Z}/(n)$ is only possible if $l \mid n$, so no other primes are possible.

This example illustrates the idea that the assassins p and q are the irreducible factors of n; the approach to primary decomposition based on modules and assassins of A/I generalises this point of view. Note that the set of zerodivisors of $\mathbb{Z}/(n)$ in $\mathbb{Z}$ is obviously just $(p) \cup (q)$; the prime factors p, q are distinguished by the fact that they are zerodivisors, but kill the smallest possible submodules of $\mathbb{Z}/(n)$.

7.4 Properties of Ass M

Proposition

(a) *Let $x \in M$ be such that $\operatorname{Ann} x = P$ is prime; then*

$$0 \neq y \in Ax \implies \operatorname{Ann} y = P.$$

In particular, $\operatorname{Ass}(A/P) = \{P\}$ for any $P \in \operatorname{Spec} A$.

(b) *Consider the set of ideals of A of the form*

$$\{\operatorname{Ann} x \mid 0 \neq x \in M\};$$

then any maximal element of this set is prime, so in $\operatorname{Ass} M$.

(c) *If A is Noetherian then $M \neq 0 \implies \operatorname{Ass} M \neq \emptyset$.*

(d) *If $L \subset M$ and $N = M/L$ then $\operatorname{Ass} M \subset \operatorname{Ass} L \cup \operatorname{Ass} N$.*

For applications, the implication (c) is the most important result of this chapter. Compare the proofs below, and Step 2 of the proof of Theorem 8.4.

Proof (a) The point is just that the submodule $Ax \subset M$ is isomorphic to A/P. Because this is an integral domain (if viewed as a ring), it follows that any nonzero element $y \in A/P$ has $\operatorname{Ann}_A y = P$.

(b) Suppose that $x \in M$ is chosen so that $P = \operatorname{Ann} x$ is maximal among annihilators of elements; I claim that P is prime. For $fg \in \operatorname{Ann} x$ means $fgx = 0$. There are two possibilities: if $gx = 0$ then $g \in P$; if $0 \neq gx \in M$ then $\operatorname{Ann}(gx) \supset \operatorname{Ann} x$, so by the assumption that $\operatorname{Ann} x$ is maximal, $\operatorname{Ann}(gx) = \operatorname{Ann} x = P$, and thus $fgx = 0$ gives $f \in P$.

(c) If A is Noetherian and $M \neq 0$, the set of ideals

$$\{ \operatorname{Ann} x \mid 0 \neq x \in M \}$$

is nonempty, and so has a maximal element, therefore $\operatorname{Ass} M \neq \emptyset$.

(d) If M contains a submodule isomorphic to A/P, there are two possible cases: either $(A/P) \cap L = 0$; then A/P maps isomorphically to a submodule of N, and hence $P \in \operatorname{Ass} N$. Or $(A/P) \cap L \neq 0$; then any $0 \neq x \in A/P \cap L$ has $\operatorname{Ann} x = P$ by (a), giving $P \in \operatorname{Ass} L$.   Q.E.D.

Corollary *If A is Noetherian and M an A-module, then*

$$\{ \text{zerodivisors for } M \text{ in } A \} = \bigcup_{P \in \operatorname{Ass} M} P.$$

Proof Because if $0 \neq m \in M$, every $a \in \operatorname{Ann} m$ is contained in an ideal $\operatorname{Ann} x$ which is maximal of this form, so is in $\operatorname{Ass} M$ (by Proposition 7.4, (b)).   Q.E.D.

Example $N = M/L$ may have assassins not in $\operatorname{Ass} M$. For example, $\operatorname{Ass}(\mathbb{Z}/(2)) = \{(2)\}$, but $(2) \notin \operatorname{Ass} \mathbb{Z}$. Thus in an exact sequence like

$$0 \to \mathbb{Z} \xrightarrow{n} \mathbb{Z} \to \mathbb{Z}/(n) \to 0,$$

the assassins of the quotient $N = \mathbb{Z}/(n)$ have nothing to do with either of the two modules $L = \mathbb{Z}$ or $M = \mathbb{Z}$, but everything to do with the special nature of the inclusion $L \hookrightarrow M$.

7.5 Relation between Supp and Ass

Theorem *Let M be a module over a ring A. Then $\operatorname{Ass} M \subset \operatorname{Supp} M$; in fact,*

$$P \in \operatorname{Ass} M \implies V(P) \subset \operatorname{Supp} M.$$

(As discussed in 5.13, $V(P)$ is the closure of P in the Zariski topology of $\operatorname{Spec} A$.)

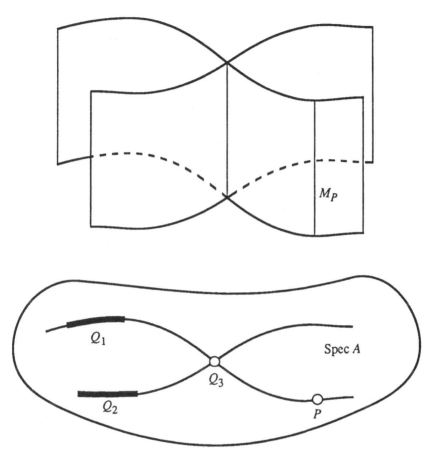

Figure 7.5. Maximal irreducible closed sets of Supp M are in Ass M; if A is Noetherian and M is finite, these are the irreducible components of Supp M.

Suppose in addition that A is Noetherian. Then a minimal element $P \in$ Supp M is in Ass M.

Recall that $P \in$ Supp M implies $\mathcal{V}(P) \subset$ Supp M. The second part of the theorem means that if $\mathcal{V}(P) \subset$ Supp M is an irreducible component (or, more generally, a maximal irreducible closed set of Supp M) then $P \in$ Ass M. See Figure 7.5

Proof By Corollary 6.7, the localisation $(A/P)_P$ of the integral domain

A/P is just the residue field $k(P)$ at P (viewed as an A-module). If $A/P \subset M$ then $0 \neq k(P) = (A/P)_P \subset M_P$, so $P \in \operatorname{Supp} M$. The same argument gives $\mathcal{V}(P) \subset \operatorname{Supp} M$: indeed, for a prime $Q \supset P$, the localisation $(A/P)_Q$ is a ring of fractions intermediate between A/P and $k(P)$, so $0 \neq (A/P)_Q \subset M_Q$, hence $Q \in \operatorname{Supp} M$.

Let $P \in \operatorname{Supp} M$ be a minimal element; then $M_P \neq 0$, but $M_Q = 0$ for any smaller prime ideal $Q \subsetneqq P$. The proof that $P \in \operatorname{Ass} M$ consists of two steps: after localising at P and viewing M_P as a A_P-module, I can prove that $\operatorname{Ass}_{A_P} M_P = \{PA_P\}$, essentially by a logical elimination. Then I have to know how to prove results by reducing to the local case, in other words, how to carry the assumptions down to the local case, and carry the conclusions back up to A.

First of all, consider M_P as a module over the localisation $B = A_P$. Then $M_P \neq 0$, so by Proposition 7.4, (c), since B is Noetherian, $\operatorname{Ass}_B M_P \neq \emptyset$; but I claim that $M_{Q'} = 0$ for any prime ideal $Q' \subset A_P$ other than the maximal ideal PA_P. Indeed, by Corollary 6.3, (iv), any prime ideal Q' of B is of the form $Q' = QB$ for a prime ideal $Q \subset P$ of A, and $(M_P)_{Q'} = M_Q$, so that by the assumption $\operatorname{Supp}_B M_P = \{PA_P\}$; that is, it has support consisting of the maximal ideal only. Therefore $\emptyset \neq \operatorname{Ass}_B M_P \subset \operatorname{Supp}_B M_P = \{PA_P\}$, and so $\operatorname{Ass}_B M_P = \{PA_P\}$.

Thus there is some $0 \neq m/s \in M_P$ such that $\operatorname{Ann}_B(m/s) = PA_P$. Now m maps to a nonzero element under the localisation map $M \to M_P$; given any two elements $t, u \notin P$, they map to units of B, so u cannot kill tm, and hence $\operatorname{Ann} tm \subset P$. The difficulty is that it might be a strictly smaller ideal. I claim that there exists $t \in A \backslash P$ such that $\operatorname{Ann}(tm) = P$.

Indeed, I can tweak m by multiplying it by some $t \in A \backslash P$ so that any given element $f \in P$ kills tm: for $(f/1) \cdot (m/s) = 0$ in M_P implies that there exists $t \in A \backslash P$ such that $ftm = tfm = 0$. Now P is finitely generated (once more using A Noetherian), so suppose $P = (f_1, \ldots, f_k)$, and let $t_i \in A \backslash P$ be such that $f_i t_i m = 0$. If I set $t = \prod t_i$ then $\operatorname{Ann}(tm)$ contains $f_1, \ldots, f_k$, hence $P = \operatorname{Ann}(tm)$.   Q.E.D.

Corollary *Let M be a finite module over a Noetherian ring A; then*

$$\operatorname{Supp} M = \bigcup_{i=1}^{n} \mathcal{V}(P_i),$$

where the P_i for $i = 1, \ldots, n$ are the finitely many minimal primes containing $\operatorname{Ann} M$; *each $P_i \in \operatorname{Ass} M$.*

Proof By Proposition 7.1, (iv), $\operatorname{Supp} M = \mathcal{V}(\operatorname{Ann} M)$; by Theorem 5.13, the set $\mathcal{V}(\operatorname{Ann} M)$ has finitely many minimal elements $\{P_1, \ldots, P_k\}$, and $\mathcal{V}(\operatorname{Ann} M) = \bigcup \mathcal{V}(P_i)$. By the theorem, these are in $\operatorname{Ass} M$. Q.E.D.

7.6 Disassembling a module

The following result, saying that any finite module over a Noetherian ring can be expressed as a successive extension of modules A/P_i with P_i prime ideals, is a key method in the study of modules. I propose the English word *disassembly* as an equivalent of the French term *dévissage* (meaning something like "unscrewing" or "taking apart") which is traditional in the subject.

Theorem *If A is a Noetherian ring and M a finite A-module, there exists a chain of submodules*

$$0 = M_0 \subset M_1 \subset \cdots \subset M_n = M$$

such that $M_i/M_{i-1} \cong A/P_i$, where $P_i \in \operatorname{Spec} A$ for $i = 1, \ldots, n$.

  Then

$$\operatorname{Ass} M \subset \{P_1, \ldots, P_n\},$$

and in particular $\operatorname{Ass} M$ is a finite set.

Proof By Proposition 7.4, (c), $\operatorname{Ass} M \neq \emptyset$, so that there exists $M_1 \subset M$ with $M_1 \cong A/P_1$. I construct the rest of the chain by the same argument applied to M/M_i: suppose that $M_0 \subset M_1 \subset \cdots \subset M_i$ are already constructed, and $M/M_i \neq 0$; then $\operatorname{Ass}(M/M_i) \neq \emptyset$, so there exists a submodule $M' \subset M/M_i$ with $M' \cong A/P_{i+1}$. Then writing $M_{i+1} \subset M$ for the inverse image of M' continues the chain. Finally, the chain must eventually stop with $M_n = M$, because M is a Noetherian module. Q.E.D.

7.7 The definition of primary ideal

I turn now from the study of modules such as A/I to the ideals themselves. An ideal Q of A is *primary* if $Q \neq A$ and for $f, g \in A$,

$$fg \in Q \implies f \in Q \text{ or } g^n \in Q \text{ for some } n > 0.$$

Equivalently, $A/Q \neq 0$, and all its zerodivisors are nilpotent. If Q is primary, set $P = \operatorname{rad} Q$; then P is prime, because

$$fg \in P \implies f^n g^n \in Q \implies \text{either } f^n \in Q \text{ or } g^{nm} \in Q$$
$$\implies \text{either } f \text{ or } g \in P.$$

Definition An ideal Q is P-*primary* (or is a primary ideal *belonging* to P) if Q is primary and $P = \operatorname{rad} Q$.

Example 1 Think of (p^n) in $\mathbb{Z}$; if I'm trying to knock an integer $g \notin (p^n)$ into (p^n) by multiplying it by f, then p must divide f.

Suppose Q is P-primary and P is finitely generated. Then

$$P^m \subset Q \subset P \text{ for some } m > 0.$$

Indeed, let $P = (f_1, \ldots, f_k)$; then $f_i^{n_i} \in Q$ for suitable $n_i \in \mathbb{N}$, since $P = \operatorname{rad} Q$. Then for $m > \sum(n_i - 1)$, every monomial of degree m in $f_1, \ldots, f_k$ is a multiple of $f_i^{n_i}$ for some i, so in Q.

Example 2 This condition is not sufficient: consider the ideal

$$I = (X^2, XY) \subset k[X, Y];$$

then $\operatorname{rad} I = (X)$, and $(X^2) \subset I \subset (X)$. But I is not primary, because $XY \in I$, but $X \notin I$ and $Y^n \notin I$ for any n. This is a favourite counterexample, and I will come back to it several times.

Lemma *If Q is an ideal such that $\operatorname{rad} Q = m$ is maximal then Q is m-primary.*

Proof Suppose $f \in A \setminus Q$, and let $I = \{g \in A \mid fg \in Q\}$; then I is an ideal of A, and $Q \subset I \subsetneq A$. So I must be contained in a maximal ideal. However, m is the only prime ideal containing Q, hence $I \subset m$. This proves

$$fg \in Q \text{ and } f \notin Q \implies g \in m = \operatorname{rad} Q,$$

so $g^n \in Q$ for some n. Q.E.D.

7.8 Primary ideals and Ass

Theorem *Let A be a Noetherian ring and Q an ideal of A; then*

$$Q \text{ is } P\text{-primary} \iff \text{Ass}(A/Q) = \{P\}.$$

The two sides are slightly different ways of saying that when multiplying $f \notin Q$ by $g \in A$, you can only get $fg \in Q$ by having $g \in P$. The proofs of each implication, although easy, use the results of the chapter so far, and need the assumption that A is Noetherian.

Proof $\implies$ Suppose Q is primary and $P = \text{rad } Q$. Then the zerodivisors of A/Q are contained in P, so that for any nonzero $x \in A/Q$

$$Q \subset \text{Ann } x \subset P = \text{rad } Q, \quad \text{and therefore} \quad \text{rad}(\text{Ann } x) = P.$$

Therefore $\text{Ann } x$ can only be a prime if it is equal to P. Moreover, $\text{Ass } A/Q \neq \emptyset$ by Proposition 7.4, (c).

$\impliedby$ The main claim is that if $\text{Ass}(A/Q) = \{P\}$ then

$$0 \neq M \subset A/Q \implies \text{rad}(\text{Ann } M) = P.$$

Indeed, by Corollary 1.12, $\text{rad}(\text{Ann } M)$ is the intersection of all prime ideals $P' \supset \text{Ann } M$. I need only take the intersection over the minimal primes $P' \supset \text{Ann } M$; these are the minimal elements of $\text{Supp } M$, so by Theorem 7.5 in $\text{Ass } M$. But $\text{Ass}(A/Q) = \{P\}$ gives $\text{Ass}(M) = \{P\}$ for every $0 \neq M \subset A/Q$. This proves the claim.

Now $Q = \text{Ann}(A/Q)$ satisfies $\text{rad } Q = P$. Let $f, g \in A$ be elements with $fg \in Q$ but $f \notin Q$. Then setting $\overline{f} = f \bmod Q \in A/Q$, I get $g \in \text{Ann } \overline{f} \subset \text{rad}(\text{Ann } \overline{f}) = P$, so $g^n \in Q$. Hence Q is a P-primary ideal. Q.E.D.

7.9 Primary decomposition

Definition Let A be a ring, $I \subset A$ an ideal; a *primary decomposition* of I is an expression

$$I = Q_1 \cap \cdots \cap Q_k \qquad (*)$$

with each Q_i primary. $(*)$ is a *shortest* primary decomposition of I if

(i) $I \subsetneqq \bigcap_{i \neq j} Q_i$ for each j (that is, no term Q_j is redundant);
(ii) Q_i is P_i-primary with $P_i \neq P_j$ for $i \neq j$.

Given any decomposition, you can easily make it shortest by ignoring the redundant terms, and grouping together the Q_i belonging to the same P, using the following obvious lemma.

Lemma *If Q_1 and Q_2 are P-primary then so is $Q_1 \cap Q_2$.*

Proof Do it yourself.

7.10 Discussion: motivation and examples

The rest of this chapter aims to prove existence and two partial uniqueness results for primary decompositions. I start by discussing the geometric motivation behind primary decomposition and giving two standard counterexamples, which show that primary components are not unique in general.

Let $A = k[X_1, \dots, X_n]$ be the polynomial ring over an algebraically closed field k, and I a radical ideal; then by the Nullstellensatz, $I = \bigcap P_i$, where $P_i = I(V_i)$ is the prime ideal corresponding to the irreducible components of the variety $X = V(I)$. Now let A be a Noetherian ring and $I \subset A$ any ideal. The idea behind primary decomposition is to express the condition $f \in I$ for an element $f \in A$ as a number of conditions $f \in Q_i$ imposed at the prime ideals P_i, where the Q_i are P_i-primary ideals. In the geometric set-up, these can usually be viewed as geometric conditions imposed along the irreducible varieties V_i, for example, requiring f to have multiplicities $\geq \mu_i$ along V_i, or imposing vanishing along V_i plus some tangency conditions on derivatives of f.

Example 1 Write $m = (X, Y) \subset k[X, Y]$ for the maximal ideal of the origin $(0,0)$. Consider the ideal $I = (X^2, XY) = m \cdot (X) \subset k[X, Y]$. Every $f \in I$ is a polynomial $f(X, Y)$ that vanishes along $X = 0$, and has multiplicity ≥ 2 at the origin. Conversely any polynomial satisfying these two conditions is a multiple gX with $g \in m$. Therefore I has the primary decomposition $I = (X) \cap (X, Y)^2$ (see Figure 7.10). The two primary components belong to the prime ideals (X) and m. Requiring $f \in (X)$ is the condition that $f(0, y) = 0$ for any sufficiently general point y of the line $X = 0$.

However, I can equally well be written as

$$I = (X^2, XY) = (X) \cap (X^2, Y - \alpha X) \qquad \text{for any } \alpha \in k.$$

The second m-primary component $(X^2, Y - \alpha X) = m^2 + (Y - \alpha X)$ is

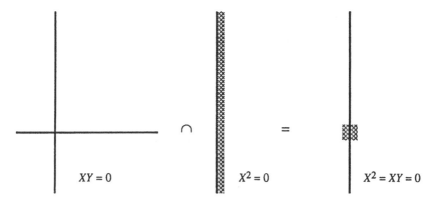

$XY = 0$ $X^2 = 0$ $X^2 = XY = 0$

Figure 7.10. The ideal $I = (X^2, XY) = (X, Y)^2 \cap (X)$ of functions vanishing on the Y-axis and at $(0,0)$ with multiplicity 2.

the ideal of polynomials that vanish at the origin, and have linear term proportional to $Y - \alpha X$. Geometrically, if you look at a curve containing $X = 0$, then requiring it to be tangent to a line $Y = \alpha X$ is equivalent to requiring it to be singular at the origin.

An important point to observe is that *along* V_i always means *at suf-ficiently general points* of V_i. If you impose a vanishing condition along V_i and some further conditions along a subvariety $W_i \subset V_i$, the ideal you get will have at least two primary components, as in Example 1. Missing out some implicit conditions holding at a smaller locus of V_i is an easy error to make.

Example 2 Let $A = k[x, y, z]/(xz - y^2)$, and $P = (x, y)$. This is the example of Figure 0.5: $A = k[V]$ is the coordinate ring of the ordinary quadric cone $V \subset k^3$ defined by $xz = y^2$, and P the ideal of functions vanishing along the line $L : (x = y = 0)$. You might reasonably expect P^2 to be the ideal of all functions vanishing twice along L, but this is not true: x vanishes twice along L, but is not in P^2. In fact x vanishes twice at every sufficiently general point of L (because of $x = y^2/z$), but the elements of P also vanish at $(0,0,0)$, so P^2 satisfies an extra condition at $(0,0,0)$. You can easily see that in A,

$$P^2 = (x^2, xy, y^2) = x \cdot (x, y, z) = (x) \cap (x, y, z)^2.$$

7.11 Existence of primary decomposition

This is the result of Emmy Noether referred to in the introduction to this chapter, and its proof is a showpiece of abstract algebra.

Theorem *In a Noetherian ring A, every ideal I has a primary decomposition.*

Definition An ideal $I \subset A$ is *indecomposable* if it cannot be written as an intersection of two strictly bigger ideals, that is, if

$$I = J \cap K \text{ with } J, K \text{ ideals } \implies I = J \text{ or } I = K.$$

For example, a prime ideal is indecomposable.

The proof of the theorem consists of the following two steps:

Step 1 In a Noetherian ring A, every ideal I is an intersection of a finite number of indecomposable ideals.

Proof This is proved by the standard Noetherian induction argument we have already seen several times: let Σ be the set of ideals not expressible as an intersection of indecomposables. I claim $\Sigma = \emptyset$. For if $\Sigma \neq \emptyset$ then Σ has a maximal element, say $I \in \Sigma$, which cannot be indecomposable, so $I = J \cap K$ with J, K strictly bigger ideals. Then by maximality $J, K \notin \Sigma$, so that they are intersections of finitely many indecomposables, hence so is I.

Step 2 In a Noetherian ring A, every indecomposable ideal Q is primary.

Proof Note that $Q \subset A$ is indecomposable $\iff 0 \subset A/Q$ is indecomposable, and the same thing for primary, so that it is enough to prove the following:

Lemma *If B is a Noetherian ring then*

$$0 \subset B \text{ is indecomposable } \implies 0 \subset B \text{ is primary.}$$

Proof Let $x, y \in B$ with $xy = 0$. Then $y \in \operatorname{Ann} x$. Consider the chain

$$\operatorname{Ann} x \subset \operatorname{Ann}(x^2) \subset \cdots \subset \operatorname{Ann}(x^n) \subset \cdots.$$

By the ascending chain condition, $\mathrm{Ann}(x^n) = \mathrm{Ann}(x^{n+1})$ for some n. Now I claim that

$$(x^n) \cap (y) = 0.$$

For if $a \in (x^n) \cap (y)$ then $ax = 0$ (since a is also a multiple of y), and on the other hand, $a = bx^n$; but then $ax = bx^{n+1}$, so $b \in \mathrm{Ann}(x^{n+1}) = \mathrm{Ann}(x^n)$, that is, $a = bx^n = 0$.

Hence if 0 is indecomposable, $xy = 0 \implies x^n = 0$ or $y = 0$, so 0 is primary. Q.E.D.

7.12 Primary decomposition and $\mathrm{Ass}(A/I)$

Theorem (1st uniqueness theorem) *Let A be Noetherian, $I \subset A$ an ideal, and let $I = \bigcap_{i=1}^{k} Q_i$ be a shortest primary decomposition, where Q_i is P_i-primary. Then*

$$\mathrm{Ass}(A/I) = \{P_1, \ldots, P_k\},$$

and in particular, the set of primes $\{P_1, \ldots, P_k\}$ is uniquely determined by I.

Proof From $I = \bigcap_{i=1}^{k} Q_i$ it follows that there is a natural diagonal inclusion map

$$A/I \hookrightarrow \bigoplus_i A/Q_i. \qquad (**)$$

Hence $\mathrm{Ass}(A/I) \subset \bigcup \mathrm{Ass}(A/Q_i) = \{P_1, \ldots, P_k\}$. On the other hand, by the irredundancy condition, for any j,

$$0 \neq N = \bigcap_{i \neq j} Q_i/I \subset A/I,$$

and of course, under the inclusion $(**)$, N maps to zero in each component A/Q_i for $i \neq j$. Hence $0 \neq N \hookrightarrow A/Q_j$. By Theorem 7.8, $\mathrm{Ass}(A/Q_j) = \{P_j\}$. Therefore A/I contains a submodule N with $\emptyset \neq \mathrm{Ass}\,N \subset \{P_j\}$, and so $P_j \in \mathrm{Ass}(A/I)$. Q.E.D.

7.13 Primary ideals and localisation

Corollary 6.3, (iv) said that on passing to a ring of fractions $A \to S^{-1}A$, a prime ideal P disjoint from S goes to a prime ideal $P' = e(P) = S^{-1}P$ of $S^{-1}A$, and $\varphi^{-1}(P') = P$. The same (rather trivial) proof based on

Proposition 6.3, (b) shows that if Q is P-primary then $Q' = S^{-1}Q$ is P'-primary and $\varphi^{-1}(Q') = Q$.

Proposition *Let A be a ring, S a multiplicative set. Let*

$$I = Q_1 \cap \cdots \cap Q_n$$

be a shortest primary decomposition of an ideal I of A, with $P_i = \operatorname{rad} Q_i$. Renumber the factors so that $S \cap P_i = \emptyset$ if $i = 1, \ldots, m$ and $S \cap P_i \neq \emptyset$ if $i = m+1, \ldots, n$. Then

$$S^{-1}I = \bigcap_{i=1}^{m} S^{-1}Q_i \quad and \quad \varphi^{-1}(S^{-1}I) = Q_1 \cap \cdots \cap Q_m.$$

Proof The first statement follows from Corollary 6.6, (ii), and the second from what I said before the proposition.   Q.E.D.

Corollary (2nd uniqueness theorem) *Let $I = \bigcap Q_i$ be a primary decomposition of I, with $P_i = \operatorname{rad} Q_i$. Suppose that P_i is a minimal element of $\{P_1, \ldots, P_n\}$. Then setting $S = A \setminus P_i$, we have*

$$Q_i = \varphi^{-1}(S^{-1}I);$$

in particular, Q_i is uniquely determined by I and P_i. So the primary component belonging to a minimal prime is uniquely determined.

Exercises to Chapter 7

7.1   If M is an A-module, show that M can be identified with a certain subset of the sections of $\mathcal{M} \to \operatorname{Spec} A$ as in 7.2, with $m \in M$ corresponding to the function $P \mapsto m/1 \in M_P$. If S is a multiplicative set, $S^{-1}M$ can be identified with a subset of partially defined sections, defined for P with $P \cap S = \emptyset$.

7.2   Let $M = \mathbb{Z}/(2) \oplus \mathbb{Z}$ viewed as a $\mathbb{Z}$-module; then (i) prove that $\operatorname{Ass} M = \{(0), (2)\}$, and (ii) show that there exist two submodules $M_1, M_2 \subset M$ such that $M_1, M_2 \cong \mathbb{Z}$ and $M = M_1 + M_2$.

7.3   If $M = M_1 + M_2$ then $\operatorname{Ass} M = \operatorname{Ass} M_1 \cup \operatorname{Ass} M_2$, true or false?

7.4   Give an example of a ring A and an ideal I which is not primary, but satisfies the condition $fg \in I \implies f^n \in I$ or $g^n \in I$ for some n. [Hint: that is, find a nonprimary ideal whose radical is prime; there's only one counterexample in primary decomposition!]

7.5   Let M be a Noetherian module and $\varphi \colon M \to M$ a homomorphism. Use the method of proof of Lemma 7.11, Step 2 to prove that $\ker \varphi^n \cap \operatorname{im} \varphi^n = \emptyset$ for some $n > 0$ (sometimes called "Fitting's lemma").

7.6   Prove Proposition 6.3, (c) and the statement at the start of 7.13.

7.7   Show that if $\varphi \colon A \to B$ is a ring homomorphism and $Q \subset B$ is a P-primary ideal then $\varphi^{-1}(Q) \subset A$ is $(\varphi^{-1}(P))$-primary.

7.8   Let $A = k[X, Y, Z]$, and let I be the ideal $(XY, X - YZ)$; find a primary decomposition $I = Q_1 \cap \cdots \cap Q_n$, and determine $P_i = \operatorname{rad} Q_i$. [Hint: to guess the result, draw the variety $V(I)$. To prove it, note that the variety $X = YZ$ is the graph of a function, so isomorphic to the (Y, Z)-plane; consider $A \to k[Y, Z]$ sending X to YZ, and use Ex. 7.7.]

7.9   In the notation of Ex. 7.8, write $B = A/I$ for the quotient ring; find elements of B annihilated by each of the primes P_i.

7.10   Let $A = k[X, Y, Z]/(XZ - Y^2)$ and $P = (X, Y)$ (as in 7.10, Example 2), and set $M = A/P^2$.

        (a) Determine $\operatorname{Ass} M$. [Hint: use 7.10, Example 2 and Theorem 7.12.]

        (b) Find elements of M annihilated by each assassin.

        (c) Find a chain of submodules $M_1 \subset \cdots \subset M_n = M$ as in Theorem 7.6.

7.11   The two examples of primary decomposition discussed in 7.10 were in $k[X, Y]$ or $k[X, Y, Z]/I$. Find similar examples in $\mathbb{Z}[Y]$ or $\mathbb{Z}[Y, Z]/I$.

8

DVRs and normal integral domains

8.1 Introduction

I gave two elementary examples of local rings in 1.14:

$$\mathbb{Z}_{(p)} = \{a/b \in \mathbb{Q} \mid p \nmid b\} \quad \text{and} \quad k[X]_{(X)} = \{f/g \in k(X) \mid X \nmid g\}.$$

These are both local integral domains with a very particular property: the maximal ideal m is generated by one element t; every element of the ring can be written $t^n \cdot u$ with u a unit. Indeed, any nonzero element of $\mathbb{Z}_{(p)}$ is of the form $p^n a/b$ with $p \nmid a$ and $p \nmid b$.

In a similar way, the ring $\mathbb{C}\{z\}$ of convergent power series

$$\mathcal{O} = \{f(z) = \sum a_i z^i \mid a_i \in \mathbb{C} \text{ and } |a_i| < 1/\rho^i \text{ for some } \rho > 0\}$$

discussed in 1.15 has the property that every $f \in \mathbb{C}\{z\}$ is of the form $f = z^n f'$, where f' is a unit.

To give an informal definition, a *discrete valuation ring* (DVR) is a unique factorisation domain A with only one prime element t. The theory of factorisation in such a ring is thus very simple: the only question is what power of t divides an element of A. It gives rise to a *valuation* $v: A \setminus 0 \to \mathbb{N}$.

DVRs are a key tool both in number theory and in geometry. In algebraic number theory, any ideal I of the ring of integers of a number field has a factorisation as a product $I = \prod P_i^{a_i}$ of prime ideals, and it is DVRs that provide the exponents a_i. In analytic geometry, meromorphic functions on a complex manifold have zeros or poles along codimension 1 submanifolds, and valuations corresponds to the geometric idea of counting the order of zero or pole. In algebraic geometry, the same idea applies to zeros or poles of rational functions.

112

8.2 Definition of DVR

Let K be a field; a *discrete valuation* of K is a surjective map

$$v\colon K \setminus \{0\} \to \mathbb{Z}$$

with the properties

(a) $v(xy) = v(x) + v(y)$;
(b) $v(x \pm y) \geq \min\{v(x), v(y)\}$ for all $x, y \in K \setminus \{0\}$.

It is traditional to use the convention that $v(0) = \infty$, so that (a) and (b) hold even if x, y or $x \pm y$ is 0. The *valuation ring* of a discrete valuation v is the subset

$$A = \{x \in K \mid v(x) \geq 0\}.$$

It is easy to check that this is a ring; a ring of this form is a *discrete valuation ring* (DVR).

It is clear from (a) that $v(x^{-1}) = -v(x)$, and that $x \in K$ is a unit of A if and only if $v(x) = 0$, so that a DVR is a local ring with maximal ideal given by

$$m = \{x \in K \mid v(x) > 0\}.$$

Let $t \in K$ be any element such that $v(t) = 1$; then since, clearly, $v(x) > 0 \implies v(x) \geq 1$, it follows that $m = (t)$, and every ideal I of A is of the form (t^n) for some n. In particular, A is a Noetherian ring. A generator t of m is called a *parameter*, *local parameter*, *uniformising element* or *prime element* of A.

8.3 A first criterion

Lemma *Let A be a Noetherian integral domain and $t \in A$ a nonunit. Then $\bigcap_{n=1}^{\infty}(t^n) = 0$.*

Proof Given $0 \neq x \in A$, either $x \notin (t)$ or $x \in (t)$, so that $x = tx'$; now repeat the argument for x'. If $x = tx' = t^2x'' = \cdots = t^nx^{(n)}$ then $(x) \subset (x') \subset \cdots \subset (x^{(n)})$, and each step is a strict inclusion (because if $(y) = (ty)$ with $y \neq 0$ then $y = aty$, and then $(at - 1)y = 0$, which in an integral domain implies that t is a unit). So the chain must stop, which means $x^{(n)} \notin (t)$ for some n, and $x \in (t^n) \setminus (t^{n+1})$.   Q.E.D.

Proposition *Let (A, m) be a local integral domain with principal maximal ideal m, say $m = (t)$ for some $t \neq 0$. Moreover, assume that $\bigcap_{n=1}^{\infty}(t^n) = 0$, which holds automatically by the preceding lemma if A is Noetherian.*

(i) *Every $0 \neq x \in A$ is of the form $x = t^n u$ with $n \geq 0$ and u a unit.*

(ii) *Define $v(x) = n$, where $x = t^n u$ as in (i), and*

$$v(x/y) = v(x) - v(y) \quad \text{for all } x/y \in K = \text{Frac } A;$$

then v is a discrete valuation of K and A its valuation ring.

(iii) *Every nonzero ideal I of A is of the form $I = (t^n)$ for some $n \geq 0$.*

Proof (i) If $x \in (t^n) \setminus (t^{n+1})$ then $x = t^n u$ with $u \notin m$, that is, u is a unit.

(ii) If $z \in K$ then $v(z) = n \in \mathbb{Z}$ if and only if $z = t^n u$, where u is a unit of A, so $v \colon K \setminus 0 \to \mathbb{Z}$ is well defined, and is a surjective homomorphism; obviously $z \in A$ if and only if $v(z) \geq 0$. If $y \neq 0$ and $v(x) \geq v(y)$ then $x/y = t^a u \in A$, where $a = v(x) - v(y) \geq 0$ and u is a unit, so that $(x \pm y)/y = t^a u \pm 1 \in A$ and $v(x \pm y) \geq v(y)$.

Note that if $v(x) > v(y)$, the same argument gives $(x \pm y)/y = t^a u \pm 1 \in A$ with $a > 0$, so that $t^a u + 1$ is a unit of A, therefore $v(x \pm y) = v(y)$.

(iii) Let $n = \min\{\nu \mid I$ contains an element $x = t^\nu u$ with u a unit$\}$. Then it is easy to see that $I = (t^n)$.   Q.E.D.

Remark If (A, m) is a Noetherian local ring and $m = (t)$, the same argument gives that either A is a DVR or $t^n = 0$. Compare the two examples in Ex. 0.10.

8.4 The Main Theorem on DVRs

Recall from Chapter 4 that an element $x \in B$ of an extension ring $A \subset B$ is *integral* over A if it satisfies a monic relation

$$x^n + a_{n-1}x^{n-1} + \cdots + a_0 = 0, \quad \text{with } a_i \in A.$$

Let A be an integral domain with field of fractions $K = \text{Frac } A$; then A is *normal* if any element $x \in K$ that is integral over A belongs to A.

Theorem *Let A be a ring; then*

$$A \text{ is a DVR} \iff \begin{cases} A \text{ is Noetherian, normal, and } \text{Spec } A = \{0, m\} \\ \text{where } m \neq 0 \text{ is a maximal ideal.} \end{cases}$$

The condition on $\operatorname{Spec} A$ means exactly that A is a local integral domain with no prime ideals intermediate between 0 and m. This can be expressed by saying that the maximal ideal m is a minimal nonzero prime, or that A is a 1-*dimensional local integral domain.*

Theorem 8.4 is the main result on DVRs. On the left-hand side, the conditions defining a DVR are very specific and restrictive. In contrast, the conditions on the right are all general abstract algebraic statements. The usefulness of the implication $\Longleftarrow$ emerges later in this chapter: starting from a fairly general ring, taking the normalisation and the localisation at a codimension 1 ($=$ minimal nonzero) prime ideal produces rings satisfying all the conditions on the right. This reduces many questions of algebra, geometry and arithmetic to DVRs.

Proof $\Longrightarrow$ is easy. We have seen that DVR implies Noetherian, and obviously

$$\mathrm{DVR} \implies \mathrm{UFD} \implies \mathrm{normal}$$

(compare 4.4 and Ex. 0.7). Also every nonzero ideal of a DVR A is of the form m^n for some n, so $\operatorname{Spec} A = \{0, m\}$.

To prove $\Longleftarrow$, I must show that $\operatorname{Spec} A = \{0, m\}$ with $m \neq 0$ and A normal implies that m is principal. The fact that A is then a DVR will follow from Proposition 8.3. Write $K = \operatorname{Frac} A$ for the field of fractions.

Step 1 $m \neq m^2$. Any ideal I of A is a finite A-module, so $mI = I$ would imply $I = 0$ by Nakayama's lemma 2.8.2.

Step 2 I choose any element $x \in m \setminus m^2$, and prove that $m = (x)$. For this, consider the A-module $M = m/(x)$, and, by contradiction, assume that $M \neq 0$. Then by Proposition 7.4, (c), $\operatorname{Ass} M \neq 0$. This means that there exists some element $0 \neq y \in m \setminus (x)$ such that $my \subset (x)$. (This is key point of the proof; note how primary decomposition comes in here.)

Step 3 Let x, y be as in Step 2, and consider $y/x \in K$. Then $y/x \notin A$ but $(y/x)m \subset A$, and is an ideal of A. There are two cases:

Case $(y/x)m = A$ This means that $yy'/x = 1$ for some $y' \in m$, but then $x = yy'$ for $y, y' \in m$, which contradicts $x \notin m^2$.

Case $(y/x)m \subset m$ In this case, I claim that y/x is integral over A. This follows by the determinant trick 2.7, just as in the proof of Proposition 4.2, (iii) $\implies$ (i): m is a finite A-module, and multiplication by y/x is a linear map $\varphi \colon m \to m$. So φ satisfies a monic relation

$$\varphi^n + a_{n-1}\varphi^{n-1} + \cdots + a_0 = 0 \qquad \text{with } a_i \in A,$$

as an endomorphism of m. But applying this to any $0 \neq z \in m$ gives

$$((y/x)^n + a_{n-1}(y/x)^{n-1} + \cdots + a_0)z = 0 \qquad \text{with } a_i \in A,$$

and since A is an integral domain and $z \neq 0$, the claim follows.

Step 4 Since y/x is integral over A, and A is normal, $y/x \in A$; thus $y \in (x)$, and this contradiction proves that $m = (x)$. Q.E.D.

8.5 General valuation rings

Definition Let A be an integral domain and $K = \operatorname{Frac} A$ its field of fractions. Then A is a *valuation ring* if

for every nonzero element $x \in K$, either x or $x^{-1} \in A$.

A partial order $>$ on a set Γ is a *total order* if for all $a, b \in \Gamma$,

exactly one of $a > b$, $a = b$ or $a < b$ holds.

An *ordered group* Γ is an Abelian group (written additively) with a total order $>$ compatible with the addition law, in the sense that $a \geq b$ and $a' > b'$ implies $a + a' > b + b'$. Examples: $\mathbb{Z}, \mathbb{Q}, \mathbb{R}$ with the usual order. $\mathbb{Z} \oplus \mathbb{Z}$ with the lex order, that is, the order defined by

$$(m, n) > (m', n') \iff m > m' \text{ or } m = m' \text{ and } n > n'.$$

In this order, $(1, 0) > (0, n)$ for every $n \in \mathbb{Z}$.

Let $K^{\times} = K \setminus 0$ and $A^{\times} = \operatorname{units}(A)$, and write $\Gamma = K^{\times}/A^{\times}$. This is an Abelian group which we write additively, that is, $[a] + [b] = [ab]$ for $a, b \in K^{\times}$. Give Γ the partial order $>$ defined by $\rho \geq 0 \iff \rho = [a]$ for $a \in A$. In other words, for $a, b \in K$,

$$[b] \geq [a] \text{ holds in the order of } \Gamma \iff b/a \in A.$$

This just amounts to the familiar idea of taking divisibility in A as a partial order: $[a] \leq [b]$ if and only if $a \mid b$.

Proposition *A is a valuation ring if and only if $>$ is a total order on $\Gamma = K^\times/A^\times$. If this holds, the quotient map $v\colon K^\times \to \Gamma$ satisfies the two conditions*

(a) *$v(xy) = v(x) + v(y)$ and*
(b) *$v(x \pm y) \geq \min\{v(x), v(y)\}$.*

Moreover, if Γ is an ordered group and $v\colon K^\times \to \Gamma$ a surjective map satisfying (a) and (b), then the subset $A = \{a \in K \mid v(a) \geq 0\} \cup \{0\}$ is a valuation ring, and $\Gamma = K^\times/A^\times$.

In this set-up, $v\colon K^\times \to \Gamma$ is called a *valuation*, and $\Gamma = K^\times/A^\times$ the *value group* of v.

Proof To say that $>$ is a total order means that for $a, b \in K$, exactly one of the three cases

$$a/b \in A \text{ but } b/a \notin A; \quad a/b \in A^\times; \quad \text{or } b/a \in A \text{ but } a/b \notin A$$

holds, which is the same thing as the definition of valuation ring. The rest is easy. Q.E.D.

8.6 Examples of general valuation rings

Just as with discrete valuation rings, it is easy to see that a valuation ring A is a local ring with maximal ideal $m = \{a \in A \mid v(a) > 0\}$. It might appear at first sight from the definition that a general valuation ring is more or less the same as a discrete valuation ring, but nothing could be further from the truth, as we see in the next result.

Theorem *Let A be a valuation ring. Then the following conditions are equivalent:*

(i) *A is Noetherian;*
(ii) *A is a DVR.*

Proof (ii) $\implies$ (i) has already been discussed. For (i) $\implies$ (ii), note that in a valuation ring A, any finitely generated ideal I is principal; for if $I = (x_1, \ldots, x_n)$, then assuming each $x_i \neq 0$, either $x_1/x_2 \in A$, so that $I = (x_2, \ldots, x_n)$, or $x_2/x_1 \in A$, ... Assume (i). Then the maximal ideal m of A is finitely generated, so principal, say $m = (t)$, and (ii) follows by Proposition 8.3. Q.E.D.

General valuation rings arise naturally from maximal conditions on subring $A \subset K$. They also play an important role in some treatments of resolution of singularities. For more details, see [A & M], Chapter 5 or [M], Chapter 4.

Example 1 If A is a DVR with $m = (t)$, construct the ring $A_\infty = \bigcup_n A(t^{1/n})$ by adjoining all the nth roots of t. It is easy to see that this is a valuation ring with value group $\Gamma = \mathbb{Q}$. The maximal ideal $m_\infty \subset A_\infty$ consists of elements divisible by some positive fractional power of t, but this is not finitely generated.

Example 2 Let $k(x, y)$ be a function field in two independent variables. Then there is a valuation v of $k(x, y)$ with values in the ordered group $\mathbb{Z} \oplus \mathbb{Z}$ with lex order such that $v(x) = (1, 0)$ and $v(y) = (0, 1)$. Since $(1, 0) > (0, n)$ for every n, the valuation ring A of v must clearly contain y and x/y^n for every n. It is easy to see that $f/g \in A$ for every $f, g \in k[x, y]$ with $g \notin (x, y)$, and it can be seen that $A = k[x, y]_{(x,y)}[x/y^n \mid n = 0, 1, \ldots]$.

One way of checking this is in terms of composite valuations: there is a discrete valuation $v_1 \colon k(x, y) \to \mathbb{Z}$ that counts the power of x dividing a rational function, that is $v_1(f/g) = n - m$, where $x^n \mid f$ and $x^m \mid g$ are the maximal powers of x dividing f and g. The valuation ring $A_1 \subset k(x, y)$ of v_1 is $k(y)[x]_{(x)}$, with maximal ideal generated by x, and residue field $k(y)$. The second component v_2 of the valuation v comes from counting the power of y dividing the residue. For more details, see Ex. 8.7 or [M], Chapter 4, Theorem 10.1.

8.7 Normal is a local condition

The next topic is the application of DVRs to give a characterisation of normal Noetherian rings. Let A be an integral domain and $K = \operatorname{Frac} A$ its field of fractions. I start with an easy characterisation of the normal property of A in local terms.

Lemma $A_P \subset K$ for each $P \in \operatorname{Spec} A$, and

$$A = \bigcap_{P \in \operatorname{Spec} A} A_P = \bigcap_{m \in \text{m-Spec} A} A_m \subset K.$$

For all P, the local ring is just the subset $A_P \subset K$ of all fractions $a/b \in K$ for $a, b \in A$ and $b \notin P$, and the intersections take place in K.

Proof For any $x \in K$, define the *ideal of denominators* of x to be

$$D(x) = \{b \in A \mid bx \in A\}$$
$$= \{b \in A \mid x = a/b \text{ for some } a \in A\} \cup \{0\}.$$

This is clearly an ideal of A. Now for $x \in K$,

$$x \notin A \iff D(x) \neq A \iff \exists m \in \text{m-Spec } A \text{ such that } D(x) \subset m$$
$$\iff \exists m \in \text{m-Spec } A \text{ such that } x \notin A_m.$$

Hence $A = \bigcap A_m$, where the intersection is over all maximal ideals m. Q.E.D.

Proposition *Equivalent conditions:*

(i) *A is normal;*
(ii) *A_P is normal for all $P \in \text{Spec } A$;*
(iii) *A_m is normal for all $m \in \text{m-Spec } A$.*

Proof (i) $\implies$ (ii) More generally, I claim that A normal implies that $S^{-1}A$ is normal for any multiplicative set S of A. For if $x \in K$ is integral over $S^{-1}A$ then

$$x^n + a_{n-1}x^{n-1} + \cdots + a_0 = 0,$$

where $a_i = b_i/c_i$ with $b_i \in A$ and $c_i \in S$. Then if I multiply this relation through by $(c_0 c_1 \cdots c_{n-1})^n$ I get an integral dependence relation for $(c_0 c_1 \cdots c_{n-1})x = y$ over A. So using the fact that A is normal, $y \in A$, and then $x = y/(c_0 c_1 \cdots c_{n-1}) \in S^{-1}A$.

(ii) $\implies$ (iii) is trivial.

(iii) $\implies$ (i) Let $x \in K$ be integral over A; then the relation

$$x^n + a_{n-1}x^{n-1} + \cdots + a_0 = 0,$$

for x over A is also a relation for x over A_m for any $m \in \text{m-Spec } A$. Therefore by the fact that A_m is normal, $x \in A_m$. So $x \in \bigcap A_m = A$. Q.E.D.

8.8 A normal ring is a DVR in codimension 1

Lemma *Let A be a normal Noetherian integral domain.*

(a) *If P is a minimal nonzero prime ideal of A then A_P is a DVR.*

(b) *Suppose that $0 \neq I = (x) \subset A$ is a principal ideal; then*

$$P \in \text{Ass } A/I \implies P \text{ is a minimal nonzero prime ideal.}$$

Proof (a) A_P is normal by Proposition 8.7, (i) $\implies$ (ii), and Noetherian by Corollary 6.3, (ii). Also by Corollary 6.3, (iv),

$$\text{Spec } A_P = \{0, PA_P\}.$$

Now PA_P is a nonzero maximal ideal, so that the result follows from Theorem 8.4.

(b) I break this up into steps:

Step 1 Reduction to a local ring. Write $(B, m) = (A_P, PA_P)$ and $I' = IB = xB$. Then B, m and I' satisfy all the assumptions of (b): by Corollary 6.7, $B/I' = A_P/IA_P = S^{-1}(A/I)$ where $S = A \backslash P$. Moreover, if $P \in \text{Ass } A/I$ then by Proposition 7.13, $m \in \text{Ass}(B/I')$. If I succeed in proving that m is a minimal nonzero prime of B then P is a minimal nonzero prime of A by Corollary 6.3, (iv); hence it is enough to prove (b) with A replaced by the local ring B and P by the maximal ideal m.

Step 2 Now $m \in \text{Ass}(B/(x))$, so that there exist $y \in B \backslash (x)$ with $my \subset (x)$; then $y/x \in K = \text{Frac } B$ satisfies $m(y/x) \subset B$. I claim that $m = (t)$ where $t = x/y$. Exactly as in the proof of Theorem 8.4, $m(y/x) \subset B$ is an ideal, so that there are two possibilities:

Case $m(y/x) \subset m$ Then by the same argument as in Theorem 8.4, Step 3, y/x is integral over B, hence in B (using the normality of B). This contradicts the choice of y.

Case $m(y/x) = B$ Then $1 \in m(y/x)$ means that $1 = t(y/x)$ for some $t \in m$; then for any $a \in m$, $a(y/x) = a/t \in B$, so that $a \in (t)$.

Step 3 Since $m = (t)$, Proposition 8.3 implies that B is a DVR, and m is a minimal nonzero prime ideal.   Q.E.D.

8.9 Geometric picture

If X is an irreducible variety and $A = k[X]$, a minimal nonzero prime P of A corresponds to a codimension 1 subvariety $Y \subset X$, that is, a hypersurface in X. If X is normal then

$$A_P = \{f/g \in k(X) \mid g \neq 0 \text{ on } Y\}$$

consists of rational functions on X defined in a neighbourhood of most points of Y. The first part of Lemma 8.8 states that this is a DVR: the valuation measures the order of zero or pole of a function along Y.

The second part of Lemma 8.8 says that, for any function $x \in k[X]$, any component of the variety $V(x) \subset X$ defined by $x = 0$ has codimension 1. In other words, if a polynomial function on a normal variety has zeros then it has zeros along hypersurfaces. The following result has a similar geometric interpretation: if X is a normal variety then any element $x \in k(X)$ is either in $k[X]$, or has poles along a codimension 1 subvariety. Thus $A = k[X] \subset k(X)$ can be characterised as the set of rational functions with no poles along hypersurfaces of X.

8.10 Intersection of DVRs

Theorem *Let A be a normal Noetherian integral domain, and K its field of fractions. Then*

$$A = \bigcap A_P,$$

where the intersection is taken over minimal nonzero primes P of A. In particular, A is an intersection of DVRs.

Proof The inclusion $\subset$ is trivial, so I only need to prove $\supset$. Let $x = b/c \in K$, and set

$$D = D(x) = \{\text{denominators of } x\} = \{a \in A \mid ax \in A\};$$

then $D = \{a \in A \mid ab \in (c)\} = \mathrm{Ann}(\bar{b})$, where $\bar{b} \in A/(c)$ is the class of b.

Suppose that $x \notin A$, or equivalently, that $\bar{b} \neq 0$. First, I claim that $D \subset P$ for some $P \in \mathrm{Ass}(A/(c))$. Indeed, $D = \mathrm{Ann}(\bar{b})$ for some $0 \neq \bar{b} \in A/(c)$. Since A is Noetherian, D is contained in a maximal ideal of this form, and by Proposition 7.4, (b), this is an element of $\mathrm{Ass}(A/(c))$.

Then by Lemma 8.8, (b), any $P \in \text{Ass}(A/(c))$ is a minimal nonzero prime; but if $D \subset P$ then $x \notin A_P$. This proves that

$$x \notin A \implies x \notin A_P \text{ for some minimal nonzero prime } P. \quad \text{Q.E.D.}$$

8.11 Finiteness of normalisation

Suppose that A is an integral domain (say, Noetherian, or finitely generated over a field) and $A \subset K = \text{Frac } A$ its field of fractions. Let $\tilde{A}$ be the integral closure of A in K (defined in 4.4).

General Question *Is $\tilde{A}$ a finite A-module?*

The answer to this is no in general, but the counterexamples, due to Akizuki and Nagata, are very complicated (see 9.4–9.5). It does hold under extra assumptions, and then $\tilde{A}$ remains Noetherian, or finitely generated over a subfield, if A was. In many cases, we can then replace the study of A by that of $\tilde{A}$.

Theorem *Let A be a normal Noetherian ring, $K = \text{Frac } A$ and $K \subset L$ a finite separable field extension. Set $B \subset L$ for the integral closure of A in L.*

Then B is a finite A-module, and in particular is a Noetherian ring.

Corollary

   (i) *Let A be an integral domain that is finitely generated over a field k; then $\tilde{A}$ is a finite A-module.*

   (ii) *Suppose that L is an algebraic number field and B the ring of integers of L; that is, L is a finite extension field $\mathbb{Q} \subset L$ with $[L : \mathbb{Q}] < \infty$ and B is the integral closure of $\mathbb{Z}$ in L. Then B is a finite $\mathbb{Z}$-module.*

The significance of this result for the resolution of singularities of algebraic curves was discussed in 4.5.

Proof (i) By the Noether normalisation lemma 4.6, I can assume that

$$k \subset k[z_1, \ldots, z_r] \subset A,$$

where $z_1, \ldots, z_r$ are algebraically independent over k and A is finite over $k[z_1, \ldots, z_r]$; I can moreover assume that $K = \text{Frac } A$ is a separable extension of $k(z_1, \ldots, z_r)$. Then $\widetilde{A}$ is the integral closure of $k[z_1, \ldots, z_r]$ in K, so that Theorem 8.11 is applicable. (ii) follows directly.   Q.E.D.

Note that the rings $\widetilde{A}$ and B in either case (i) or (ii) are normal Noetherian rings, so by Theorem 8.10 are intersections of DVRs. This allows us to use DVRs as a systematic method in both algebraic geometry and algebraic number theory.

8.12 Proof of Theorem 8.11

The aim is to express B as a submodule of a finite A-module

$$B \subset \sum_{i=1}^{n} Ax_i \subset L,$$

and then to conclude by Corollary 3.5, (iii).

Without loss of generality, I can assume that L/K is a Galois extension. (Indeed, if not, let L_1 be the Galois closure of L/K and B_1 the integral closure of A in L_1; if B_1 is a finite A-module, the same holds for B.)

If L/K is a Galois extension, with $G = \mathrm{Gal}(L/K)$, then set

$$\mathrm{Tr}(x) = \sum_{g \in G} g(x) \in K \qquad \text{for } x \in L.$$

(See the appendix 8.13 for more details).

Proposition *Let L/K be a Galois extension.*

(a) *If $b \in B$ then $\mathrm{Tr}(b) \in A$.*

(b) *The map $(x, y) \mapsto \mathrm{Tr}(xy)$ defines a nondegenerate bilinear form $L \times L \to K$.*

Proof of (a) Consider the polynomial $\Phi(x) = \prod_{g \in G}(x - g(b))$, which is invariant under G. Therefore all its coefficients are elements of K. On the other hand $b \in B$ is integral over A, hence so is $g(b)$ for every $g \in G$. Therefore all the coefficients of $\Phi(x)$ are integral over A, hence in A, and $\mathrm{Tr}(b) \in A$, because it is one of the coefficients of $\Phi(x)$.

For the proof of (b), see Theorem 8.13 below or [M], Chapter 9, §26.

Proof of Theorem 8.11 Choose elements $y_1, \ldots, y_n \in B$ forming an L-basis of K. Then since $\mathrm{Tr}(xy)$ is a nondegenerate bilinear form, there exists a dual basis $x_1, \ldots, x_n \in L$ of L/K such that

$$\mathrm{Tr}(x_i y_j) = \delta_{ij} \qquad \text{(Kronecker delta)}.$$

Then

$$\sum A y_i \subset B \implies B \subset \sum A x_i, \qquad (*)$$

and therefore B is contained in a finite A-module. To prove $(*)$, note that any $b \in B$ can be written $b = \sum c_i x_i$ with $c_i \in K$, since the x_i form a basis of L/K. But then for each j, we know $\mathrm{Tr}(b y_j) \in A$ by the Proposition, (a) and on the other hand, since $\mathrm{Tr}(x_i y_j) = \delta_{ij}$, I have

$$\mathrm{Tr}(b y_j) = \sum_j c_i \, \mathrm{Tr}(x_i y_j) = c_j. \qquad \text{Q.E.D.}$$

8.13 Appendix: Trace and separability

This appendix is pure Galois theory, and is provided to make the proof of Theorem 8.11 self-contained.

Definition of the trace If $K \subset L$ is a finite field extension then the trace map $\mathrm{Tr}_{L/K} \colon L \to K$ is defined as follows: for $\alpha \in L$, write

$$f_\alpha(X) = X^d + c_1 X^{d-1} + \cdots + c_{d-1} X + c_d \in K[X]$$

for the minimal polynomial of α (so that the primitive extension generated by α is $K(\alpha) \cong K[X]/(f_\alpha)$), and set

$$\mathrm{Tr}_{L/K}(\alpha) = -[L : K(\alpha)]c_1.$$

If f_α splits as $f_\alpha(X) = \prod_{i=1}^{d}(X - \alpha_i)$ over some extension field of K then the coefficients of f_α are $\pm$ the elementary symmetric functions of the roots $\alpha_1, \ldots, \alpha_d$, and in particular $-c_1 = \sum \alpha_i$; hence $\mathrm{Tr}_{L/K}(\alpha) = [L : K(\alpha)] \cdot \sum \alpha_i$. In the special case that $K \subset L$ is Galois with group $G = \mathrm{Gal}(L/K)$ then $\mathrm{Tr}_{L/K}(\alpha) = \sum_{g \in G} g(\alpha)$.

Remark An alternative definition of the trace map, which I do not need here, applies to any finite K-algebra L, not necessarily commutative: let $\mu_\alpha \colon L \to L$ be the multiplication map $y \mapsto \alpha y$, and set $\mathrm{Tr}_{L/K}(\alpha) = \mathrm{Tr}(\mu_\alpha)$, the trace in the sense of linear algebra.

It is easy to see that $\text{Tr}_{L/K}\colon L \to K$ is K-linear, and is either zero or surjective (because the image K is 1-dimensional).

Lemma *Let $K \subset L = K(\alpha)$ be a primitive field extension of degree d, and suppose that $\text{Tr}_{L/K}(\alpha^k) = 0$ for $k = 1, \ldots, d$. Then α is inseparable, $p \mid d$ and $\text{Tr}_{L/K} = 0$.*

Proof Suppose that α is separable. Then the minimal polynomial of α has d distinct roots $\alpha_1, \ldots, \alpha_d$ in some big enough field extension, which are the conjugates of α. Then the conjugates of α^k are $\alpha_1^k, \ldots, \alpha_d^k$, and $\text{Tr}_{L/K}(\alpha^k)$ is the sums of powers $\text{Tr}_{L/K}(\alpha^k) = \Sigma_k = \sum \alpha_i^k$. The assumption of the lemma is that $\Sigma_k = 0$ for $k = 1, \ldots, d$.

Now there is a famous formula called Newton's rule that expresses the sums of powers Σ_k recursively in terms of the elementary symmetric functions $\sigma_i = (-1)^i c_i$:

$$\Sigma_k - \sigma_1 \Sigma_{k-1} + \sigma_2 \Sigma_{k-2} - \cdots + (-1)^{k-1} \sigma_{k-1} \Sigma_1 + (-1)^k k \sigma_k = 0$$

(for $k \leq d$).

Since $\Sigma_k = 0$, this implies that $k c_k = 0$ for $k = 1, \ldots, d$. However, $c_d \neq 0$, since f_α is an irreducible polynomial, and therefore $d = 0$ in k, that is, $p = \text{char } K > 0$ and $p \mid d$. Hence also $(d-k)c_k = 0$ for $k = 0, \ldots, d$, that is, f_α is a polynomial in X^p only, as required.   Q.E.D.

Theorem *Let $K \subset L$ be a separable finite field extension. Then*

$$(x, y) \mapsto \text{Tr}_{L/K}(xy)$$

defines a nondegenerate symmetric bilinear form $L \times L \to K$ over K.

Proof The fact that $\text{Tr}_{L/K}(xy)$ is bilinear and symmetric in x and y follows obviously, because $\text{Tr}_{L/K}$ is K-linear. To prove it is nondegenerate, I assume that some element $\alpha \in L$ satisfies $\text{Tr}_{L/K}(\alpha y) = 0$ for every $y \in L$ and prove that $K \subset L$ is inseparable.

It is an exercise to see that if $K \subset K_1 \subset L$ is an intermediate field, then

$$\text{Tr}_{L/K} = \text{Tr}_{K_1/K} \circ \text{Tr}_{L/K_1}.$$

Consider the intermediate field $K \subset K(\alpha) \subset L$; then for all y,

$$\begin{aligned}
0 = \text{Tr}_{L/K}(\alpha y) &= \text{Tr}_{K(\alpha)/K}\left(\text{Tr}_{L/K(\alpha)}(\alpha y)\right) \\
&= \text{Tr}_{K(\alpha)/K}\left(\alpha \, \text{Tr}_{L/K(\alpha)}(y)\right).
\end{aligned}$$

Now if $\mathrm{Tr}_{L/K(\alpha)} = 0$, since $[L : K(\alpha)] < [L : K]$, it follows by induction that L is inseparable over $K(\alpha)$, hence also over K. The alternative is that $\mathrm{Tr}_{L/K(\alpha)}$ is surjective to $K(\alpha)$. In this case the expression in the inner bracket is an arbitrary element of $K(\alpha)$, so that $\mathrm{Tr}_{K(\alpha)/K}(\alpha^k) = 0$ for all k, and the result follows by the lemma.   Q.E.D.

Exercises to Chapter 8

8.1   Using Nakayama's lemma, show that if (A, m) is a Noetherian local ring then the maximal ideal m is principal if and only if m/m^2 is 1-dimensional over $k = A/m$. Deduce from Proposition 8.3 that A is a DVR if and only if A is Noetherian, local with $\mathrm{Spec}\, A = \{0, m\}$, and m/m^2 is 1-dimensional over $k = A/m$.

8.2   **DVRs and nonsingular curves** Let k be an algebraically closed field and $f \in k[X, Y]$ an irreducible nonconstant polynomial of the form

$$f = l(X, Y) + g(X, Y)$$

with $l(X, Y) = aX + bY$ and $g \in (X, Y)^2$; set $R = k[X, Y]/(f)$ and $P = (X, Y)/(f)$, and $(A, m) = (R_P, m_P)$. Prove that A is a DVR if and only if $l \neq 0$. [Hint: use Ex. 8.1.]

   The result says that A is a DVR if and only if the plane curve $C : (f = 0) \subset k^2$ is nonsingular at $(0, 0)$.

8.3   **Completion of DVR** Let A be a DVR with parameter t. A sequence $s_0, s_1, \ldots$ of elements of A is a *t-adic Cauchy sequence* if

$$(\forall i) \quad (\exists n_0) \quad \text{such that} \quad (\forall n, m \geq n_0) \quad s_n - s_m \in (t^i).$$

Introduce an equivalence relation on Cauchy sequences (the "obvious" one), and define addition and multiplication operations to make classes of Cauchy sequences into a ring $\widehat{A}$, the *completion* of A. Prove that $\widehat{A}$ is a DVR with the same parameter t, and that $A/t^n A = \widehat{A}/t^n \widehat{A}$ for all $n \geq 0$.

   Show that the rings $k[\![x]\!]$ and $\mathbb{Z}_p$ of Ex. 0.12 are the respective completions of the DVRs $k[x]_{(x)}$ and $\mathbb{Z}_{(p)}$ of 1.14.

   An equivalence class of Cauchy sequences s_i can be written (nonuniquely, of course) as a power series in the parameter t:

$$z = a_0 + a_1 t^{n_1} + a_2 t^{n_2} + \cdots,$$

with $a_i \in A$ and $0 < n_1 < n_2 < \cdots$.

8.4   Let A be an intermediate ring $k[x] \subset A \subset k[[x]]$ (ring of formal
        power series, see Ex. 0.12), and suppose that A is local with
        maximal ideal (x). Prove that A is a DVR, and in particular A
        is Noetherian. [Hint: use Proposition 8.3.]

        Local rings of this type include the localisation of very com-
        plicated rings, but they are all covered by the theory of DVRs.
        They can be viewed from an analytic point of view as rings of
        power series with curious convergence conditions.

8.5   Let A be a DVR and $z_0 = z = a_0 + a_1 t^{n_1} + a_2 t^{n_2} + \cdots \in \widehat{A}$ as
        in Ex. 8.3. Define elements $z_r \in \widehat{A}$ by

        $$z_r = \frac{z_0 - \text{first } r \text{ terms}}{t^{n_r}} = a_r + a_{r+1} t^{m_{r+1}} + \cdots, \quad (1)$$

        where $m_{r+1} = n_{r+1} - n_r$; observe the identities

        $$z_r - a_r = t^{m_{r+1}} z_{r+1}, \quad (2)$$

        and set $B = A[z_0, z_1, \ldots] \subset \widehat{A}$. Prove that there is a ring
        homomorphism $B \to k = A/(t)$ sending $t \mapsto 0$ and $z_r \mapsto a_r$,
        with kernel the principal ideal (t).

        Therefore (t) is a maximal ideal. Prove that the localisation
        $B_{(t)}$ is a DVR. [Hint: Argue exactly as in Ex. 8.4.]

        The interesting case of this construction is when z_0 is trans-
        cendental over A. (For DVRs of interest, $\widehat{A}$ usually has infi-
        nite transcendence degree over A.) Then $A \subset B_{(t)}$ is an ex-
        tension of DVRs with the same parameter t, but with $B_{(t)}$ of
        transcendence degree 1 over A. For example if $A = \mathbb{C}[t]$ and
        z_0 has positive radius of convergence, consider the analytic arc
        $C : (z = z_0(t)) \subset \mathbb{C}^2$. Then $B_{(t)}$ is the ring of regular functions
        near 0 on C that are rational functions in z, t.

8.6   Let A be a general valuation ring, m its maximal ideal and
        $P \subset m$ another prime ideal. Prove that there is a valuation ring
        $B \supset A$ such that P is the maximal ideal of B; and A/P is a
        valuation ring of the field B/P.

8.7   Let $C = k[Y]_{(Y)}$; this is a valuation ring of $k(Y)$. Set $K = k(Y)$
        and let $B = K[X]_{(X)}$; this is a valuation ring of $K[X]$; finally,
        let $A \subset B$ be the inverse image of C under the surjection $B \twoheadrightarrow
        K = B/(X)$. Prove that A is a valuation ring whose value group
        is $K^*/A^* \cong \mathbb{Z} \oplus \mathbb{Z}$ with lex order.

8.8   Exercises to Appendix 8.13: (i) Prove Newton's rule. (ii) Prove
        that $\mathrm{Tr}_{L/K} = \mathrm{Tr}_{K_1/K} \circ \mathrm{Tr}_{L/K_1}$ for an intermediate field $K \subset$

$K_1 \subset L$. (iii) Prove that a finite extension $K \subset L$ is inseparable if and only if $\mathrm{Tr}_{L/K} = 0$.

9
Goodbye!

This final chapter starts with an overview, relating the material of the book to more advanced subjects. There is a brief discussion of extra topics that I would like to have included, but that would have stretched the material beyond what can be covered in a 30 hour undergraduate lecture course. In 9.4–9.5 I treat some counterexamples, for example to finiteness of normalisation, that are traditionally regarded as "difficult". Apart from the evangelical aspiration of making these examples accessible to a wider public, my purpose here is to illustrate some ways in which Noetherian rings can be more general than geometric rings $k[V]$, while retaining a strong geometric flavour.

I also discuss in 9.7–9.9 the current sociological status of algebra in math. With research experience restricted to algebraic geometry, I do not claim any particular authority to write on algebra, but it is clear that there is room in the subject for views other than those of professional algebraists.

9.1 Where we've come from

Chapters 1–4 and 6 discuss fairly harmless general ideas in algebra such as generators of a module and Noetherian rings. The material is standard, but I have mixed in some more substantial things like the picture of Spec $\mathbb{Z}[X, Y]$ in Chapter 1 and the discussion of resolution of singularities of curves in Chapter 4, partly in the hope of keeping the reader awake.

Chapters 5, 7 and 8 are three substantial individual topics, and the undergraduate exam always contains questions on at least two of these three chapters. Chapter 5 shows how to view a geometric ring A as the coordinate ring $A = k[V]$ of an algebraic variety; it proves that

Spec A is the set of irreducible subvarieties $X \subset V$, and introduces the Zariski topology on Spec A both in the context of the coordinate ring $k[V]$, and for general rings. The Zariski topology on an affine variety is a more or less tautological definition, but it has two really important applications. The first is in algebraic geometry, where more general varieties (for example, projective varieties) arise by glueing affine varieties together along Zariski open sets; the key point here is that a rational function $f \in k(V) = \operatorname{Frac} k[V]$ is regular on a dense Zariski open set of V (namely, where a denominator of f is nonzero), so that we know what local functions there should be on the glued-up variety. (See, for example, [Sh1] for a proper treatment.) The second is the generalisation to the spectrum Spec A of a general ring, which makes the intuition arising from the geometric case applicable in a very wide context. It is no exaggeration to say that this analogy provides the inspiration for the whole of commutative algebra, and the driving force behind many of its successes.

Chapter 7 discusses how the language of the geometry of Spec A applies to ideals and modules, and how this relates to the classical and Bourbakist notions of primary decomposition. As I discuss in more detail in 9.3, the material of Chapter 8 on DVRs and normal rings lays the foundations for working with normal varieties and their divisors and functions.

9.2 Where to go from here

This course has been used several times as an M.Sc. course together with extra reading, made up from the following topics covered in [A & M], Chapters 10–11.

(i) I-adic completion of rings For example, the ring of formal power series $k[\![X_1, \ldots, X_n]\!]$ can be viewed in terms of Taylor series of polynomials at the origin up to any order. In other words,

$$k[\![X_1, \ldots, X_n]\!] = \varprojlim k[X_1, \ldots, X_n]/m^k,$$

where $m = (X_1, \ldots, X_n)$ is the maximal ideal, and $k[\underline{X}]/m^k$ the ring of Taylor series up to order $k - 1$. For many purposes of algebra and algebraic geometry, formal power series expansions are a reasonable substitute for the convergent power series of complex analysis (see for example [Sh1], Chapter II, 2.2). The p-adic completion of a number field at a prime plays a similar fundamental role in number theory (see for example [B & Sh], Chapter 1, §§3–6 and Chapter 4).

(ii) Dimension theory of local rings Dimension and codimension have appeared subliminally several times in the text. There are four completely different treatments of dimension in algebraic geometry and commutative algebra. Most intuitive are the two definitions of dimension of an algebraic variety discussed in [UAG], §6 and [Sh1], Chapter I, §6 and Chapter II, 1.4 in terms of the transcendence degree of the function field, and the dimension of the tangent space at a general point. Next is the *Krull dimension* of a ring A, defined as the maximum length of chains of prime ideals

$$P_0 \subset P_1 \subset \cdots \subset P_n \subset A.$$

Compare the remark after Theorem 8.4. By the Nullstellensatz correspondence Corollary 5.8, if $A = k[V]$, this just boils down to the maximum length of chains of irreducible subvarieties

$$V \supset X_0 \supset X_1 \supset \cdots \supset X_n.$$

(Compare [Sh1], Chapter I, 6.2.)

The fourth definition is the least intuitive, but technically the most powerful and far-reaching. It is in terms of Hilbert functions and their asymptotic behaviour: the idea, roughly, is that the polynomial ring $k[X_1, \ldots, X_n]$ has $\binom{n+r}{n}$ monomials of degree $\leq r$, that is,

$$\dim_k k[X_1, \ldots, X_n]/m^{r+1} = \frac{1}{n!}r^n + \text{(lower order terms)}.$$

Equivalently, $\dim m^r/m^{r+1} \sim (1/(n-1)!)r^{n-1}$ as $n \to \infty$. For a general Noetherian local ring A, m, note that m^r/m^{r+1} is a finite dimensional vector space over $k = A/m$ (even if A does not contain a field). Thus $\dim A$ can be measured in terms of the asymptotic growth as $r \to \infty$ of $\dim m^r/m^{r+1}$. Working out this idea leads to the professional's definition of dimension, as well as spin-offs in a number of contexts, such as multiplicities and intersection numbers.

(iii) Regular local rings We've seen that a DVR is a 1-dimensional local ring A, m with principal maximal ideal $m = (t)$. In general, one proves that for an n-dimensional local ring A, m, the maximal ideal needs $\geq n$ generators. Equality here is the defining condition of a *regular local ring*; elements t_i such that $m = (t_1, \ldots, t_n)$ are *regular local parameters* of m. A regular local ring enjoys good properties analogous to DVRs. For example, in the geometric case (over an algebraically closed field), the local ring $\mathcal{O}_{X,P}$ is regular if and only if $P \in X$ is a nonsingular point, and then $(t_1, \ldots, t_n)$ are analogues of analytic coordinates on an n-dimensional manifold. In particular, elements of $\mathcal{O}_{X,P}$ have Taylor expansions in the $(t_1, \ldots, t_n)$, that is, $\mathcal{O}_{X,P}/m_P^r \cong k[t_1, \ldots, t_n]/m^r$ for any r.

9.3 Tidying up some loose ends

I gather here a number of useful corollaries and applications. Most are fairly substantial foundational results that follow directly from the results in the main text, but which did not fit in there for lack of space or the appropriate language. For simplicity, I assume in what follows that varieties are defined over an algebraically closed field.

(a) Branched covers and Noether normalisation In algebraic geometry, there is a one-to-one correspondence between polynomial maps of varieties $f: V \to W$ and ring homomorphisms $\varphi: k[W] \to k[V]$ between their coordinate rings (see, for example, [UAG], Theorem 4.5 or [Sh1], Chapter I, 2.3). If $k[V]$ is the coordinate ring of a variety V, an inclusion map $k[x_1, \ldots, x_n] \subset k[V]$ corresponds to the polynomial map $\pi: V \to k^n$ defined by $x_1, \ldots, x_n$.

A particular case is the linear projection $\pi: V(F) \to k^n$ of a hypersurface $V(F) \subset k^{n+1}$ defined by $x_1, \ldots, x_n$. This was mentioned at the start of 0.5, where I said cryptically that you can *almost always* solve $F = 0$ for x_0 in terms of $x_1, \ldots, x_n$. You can think of $\pi: V(F) \to k^n$ as a branched cover, but it is not finite in general: it may happen that above some particular value $P = (a_1, \ldots, a_n) \in k^n$, the equation $F(x_0, a_1, \ldots, a_n)$ is identically equal to 0 (so that $f^{-1}P$ is the whole X-line), or a nonzero constant (so that $\pi^{-1}(P) = \emptyset$).

Noether normalisation (Theorem 4.6) proves that the coordinate ring of a variety $k[V]$ can be expressed as a finite extension of a polynomial ring. If $k[y_1, \ldots, y_n] \subset k[V]$ is finite, the corresponding projection map $\pi: V \to k^n$ defined by $y_1, \ldots, y_n$ is a *finite branched cover*: every point $P \in k^n$ has a nonempty finite set of inverse images. This was proved in Ex. 4.12.

The geometric meaning of finite is as follows: if $x \in k[V]$ is integral over $k[y_1, \ldots, y_n]$ with monic relation

$$F(x, \underline{y}) = x^n + a_{n-1}x^{n-1} + \cdots + a_1x + a_0 = 0, \tag{1}$$

then (1) determines a finite nonempty set of values of x for every $P = (y_1, \ldots, y_n) \in k^n$. In fact, since F is monic and its coefficients a_i are polynomials in $\underline{y}$, you can view the set of roots of (1) as depending continuously on $\underline{y}$. When $k = \mathbb{C}$ (with the classical topology), it is easy to bound the roots of (1) in terms of $(a_1, \ldots, a_n)$. Then when $\underline{y}$ runs through a bounded set in $\mathbb{C}^n$, the set of roots of (1) are also bounded. Therefore $\pi\colon V \to \mathbb{C}^n$ has the property that the inverse image of a compact set is compact, or in other words, π is *proper*. Slightly less precisely, we express this by saying that $\pi\colon V \to \mathbb{C}^n$ does not lose solutions going off to infinity. In fact it can be proved that the number of points of $\pi^{-1}(P)$ is constant over a dense Zariski open set of k^n, and drops over a closed set, the *branch locus* of π. See for example [Sh1], Chapter II, 6.3.

(b) DVRs and nonsingular algebraic curves If $C \subset k^n$ is an algebraic curve and $P \in C$ a point then, by definition, C is *nonsingular* at P if and only if the tangent line is well defined (compare [UAG], §7 or [Sh1], Chapter I). It follows at once from the simple criterion Proposition 8.3 and from Nakayama's lemma (as in Ex. 8.1) that $P \in C$ is nonsingular if and only if the local ring $\mathcal{O}_{C,P}$ is a DVR (compare also Ex. 8.2).

Next, since $\mathcal{O}_{C,P}$ is a 1-dimensional Noetherian local ring, the main result on DVRs, Theorem 8.4, says that $\mathcal{O}_{C,P}$ is a DVR if and only if it is normal. Finally, the "normal is a local condition" result Proposition 8.7 gives that $k[C]$ is normal if and only if all its local rings $\mathcal{O}_{C,P}$ are. The conclusion is that C is a nonsingular curve if and only if $k[C]$ is normal.

(c) Resolution of singularities of algebraic curves If $C \subset k^n$ is an irreducible curve, a *resolution of singularities* of C is a polynomial map $\pi\colon \widetilde{C} \to C$, where $\widetilde{C}$ is a nonsingular curve, and the map π is finite, and an isomorphism outside the singularities of C.

Finiteness of normalisation (Theorem 8.11) provides the resolution of singularities of curves in an automatic way: indeed, if C has coordinate ring $k[C] = k[x_1, \ldots, x_n]/I_C$, let $k[\widetilde{C}]$ be its normalisation (that is, the integral closure of $k[C]$ in the field of fractions $k(C)$, compare 4.4). Then by Theorem 8.11, $k[\widetilde{C}]$ is again a finitely generated algebra over k, so corresponds to a curve $\widetilde{C}$ with a finite polynomial map $\widetilde{C} \to C$.

Moreover, the coordinate ring $k[\widetilde{C}]$ is normal, so that $\widetilde{C}$ is nonsingular by what I just said.

For a completely different approach at a fairly elementary level to resolution of curve singularities, see [F].

(d) Projective varieties and normalisation Finiteness of normalisation is also a key technical result for working with projective varieties. A *projective variety* V is something isomorphic to a closed subvariety of some $\mathbb{P}^N$; it has an open cover by affine varieties V' having coordinate rings $k[V']$ to which the ideas of Chapter 5 are applicable.

As a typical example, if C is a nonsingular projective curve and $f\colon C \to \mathbb{P}^n$ a nonconstant map, the image $B = f(C)$ is an irreducible curve, and C is the normalisation of B in the extension of function fields $k(B) \subset k(C)$. It follows from this, first, that the nonsingular curve C is uniquely determined by B and the field extension $k(B) \subset k(C)$ (that is, C and the cover $C \to B$ is uniquely determined by any open affine piece of C). Secondly, the morphism $f\colon C \to B$ is finite, in the sense that over any affine piece $B' \subset B$ the inverse image $C' = f^{-1}(B')$ is an affine piece of C, and the coordinate ring $k[C']$ is finite over $k[B']$.

This result is in many cases an easy alternative to Serre's finiteness theorem on the direct image of a coherent sheaf. Among its straightforward applications are the theorem that the only regular functions on a projective variety are constant, and the theorem that for a morphism $f\colon C \to B$ between nonsingular curves the number of inverse images $f^{-1}(b)$ (with multiplicities) is constant, equal to $[k[C] : k[B]]$ (see, for example, [Sh1], Chapter III, 2.1, Theorem 1).

(e) Dedekind domains and unique factorisation Let A be the ring of integers $A = \mathcal{O}_K$ of a number field K, or the coordinate ring $A = k[C]$ of a nonsingular algebraic curve C. Then, as mentioned in 0.8, (a), A is not necessarily a UFD. However, it has unique factorisation of ideals into prime *ideals*, a fact that is the origin of the term ideal (short for *ideal number*).

This follows easily from the results of Chapters 7–8. Clearly, in either case, A is a Noetherian integral domain that is 1-dimensional (every nonzero prime ideal P is maximal) and normal. A ring satisfying these conditions is called a *Dedekind domain*. Thus every localisation A_P is a DVR by the main theorem on DVRs, Theorem 8.4. If $0 \neq I$ is an ideal of A then its localisation at P is an ideal of A_P, therefore is of the form $(t_P^n) \subset A_P$ for some $n \geq 0$. There are only finitely many prime ideals

$\{P_i\}$ for which $n_i \neq 0$; these are the elements of Ass A/I, or the *prime divisors* of I. Thus

$$I = \bigcap_i P_i^{n_i} = \prod_i P_i^{n_i}.$$

For more details, consult [M], Chapter 4, §11, [A & M], Chapter 9 or [B & Sh], Chapter 3, §6.

(f) Divisors on a normal variety and the class group In contrast to the case of curves, there is no easy way of getting a resolution of singularities in dimension ≥ 2; the definition of resolution itself is beyond the scope of this book. However, the normalisation $\widetilde{X}$ of a variety X is a normal variety, and is therefore nonsingular in codimension 1, so that the ideas of Chapter 8 apply to it. As mentioned in 8.9, the theory of DVRs gives a precise meaning to zeros and poles of rational functions: a rational function $f = g/h \in k(X)$ is regular and invertible where $g, h \neq 0$, which is the complement of a Zariski closed set $S(f)$. For each codimension 1 subvariety $\Gamma \subset S(f)$ (there are only finitely many), the local ring $k[X]_{I(\Gamma)} = \mathcal{O}_{X,\Gamma}$ is a DVR, and the divisor of f is defined to be

$$\operatorname{div} f = \sum_{\Gamma \subset S(f)} v_\Gamma(f) = \text{zeros of } f - \text{poles of } f.$$

There is no harm in extending the sum to all codimension 1 subvarieties $\Gamma \subset X$, since $v_\Gamma(f) = 0$ if Γ is not a component of $S(f)$.

On a normal variety X, a finite sum $\sum_i n_i \Gamma_i$ of codimension 1 irreducible subvarieties $\Gamma_i \subset X$ is called a *divisor*. Divisors are important in discussing functions on global varieties, for example, the Riemann–Roch theorem on a nonsingular projective curve (see [F] or [Sh1], Chapter III, 1.5 and 6.6). However, divisors also provide local invariants of singular affine varieties (or, more generally, of normal rings). The *divisor class group* Cl X is the group of divisors modulo the divisors of rational functions. It is analogous on the one hand to the ideal class group of algebraic number theory, and on the other to a homology group of a topological space. An example was discussed in 0.5 in terms of the relation between unique factorisation and the question of whether the ideal of a codimension 1 subvariety is generated by one element.

9.4 Noetherian is not enough

It may seem disappointing that, after all the success obtained using only the ring axioms and the a.c.c., further progress on a number of questions depends on assumptions of a concrete nature, so that the heart's desire of the algebraist mentioned at the start of Chapter 3 has to remain forever unrequited. Grothendieck (in [EGA], IV_2, 7.8) has developed a theory of "excellent rings" (following Akizuki, Zariski and Nagata), that assembles everything you might ever need as a list of extra axioms, but it seems that this will always remain an obscure appendix in the final chapter of commutative algebra textbooks: "Le lecteur notera que les résultats les plus delicats du §7 ne serviront qu'assez peu dans la suite".

The catch-phrase "counterexamples due to Akizuki, Nagata, Zariski, etc. are too difficult to treat here" when discussing questions such as Krull dimension and chain conditions for prime ideals, and finiteness of normalisation is a time-honoured tradition in commutative algebra textbooks (comparable to the use of fascist letters $\mathfrak{P}$ and $\mathfrak{m}$, etc., for prime and maximal ideals). This does little to stimulate enthusiasm for the subject, and only discourages the reader in an already obscure literature; I discuss here three counterexamples (taken, with some simplifications, from the famous "unreadable" appendix to [Nagata]) to show some of the ideas involved. The following section treats the first and most famous counterexample of all, due to Akizuki. All the rings here are integral domains over a field k, but, of course, not finitely generated over it.

(1) Counterexample to finiteness of normalisation Let $k \subset K$ be an infinite algebraic field extension, and consider the formal power series rings $k[\![x]\!]$ and $K[\![x]\!]$ (see Ex. 0.12), and the intermediate subring

$$k[\![x]\!] \subset A \subset K[\![x]\!]$$

defined by $f = \sum a_i x^i \in A$ if and only if the field $k(a_0, a_1, \dots)$ generated by its coefficients is finite over k; you can think of the condition $[k(a_0, a_1, \dots) : k] < \infty$ as a convergence requirement on f. In other words,

$$A = \bigcup_{[k':k]<\infty} k'[\![x]\!] \subset K[\![x]\!],$$

where the union takes place over all finite field extensions $k \subset k'$ contained in K. Alternatively, one sees easily that $A = k[\![x]\!][K] \subset K[\![x]\!]$ or $A = k[\![x]\!] \otimes_k K$. Now A is a local ring with principal maximal ideal $m = (x)$ and $\bigcap_{n \geq 0} m^n = 0$, hence is a DVR by Lemma 8.3.

Now suppose, *if possible*, that $z = z_0 = \sum b_i x^i \in K[\![x]\!]$ is a power series which is integral over A but not in A, that is, $[k(b_0, b_1, \dots) : k] =$

∞. Write

$$z_n = \sum_{i \geq n} b_i x^{i-n} = \frac{z - \text{first } n \text{ terms}}{x^n} \quad \text{for } n \geq 0.$$

Then obviously, by the formula just given, $z_n \in \operatorname{Frac} A[z]$. Suppose, *if possible*, that all the z_n are also integral over A. Then $A[z]$ is a counterexample to finiteness of normalisation. Indeed, if finite, $A[z]$ would be a Noetherian A-module, so that the ascending chain

$$A z_0 \subset A z_0 + A z_1 \subset \cdots \subset \sum_{i=0}^{n} A z_i$$

would have to stop. Hence the z_j would be A-linear combinations of $z_0, \ldots, z_n$. But then the leading coefficients b_j of all the z_j would have to be k-linear combinations of $b_0, \ldots, b_n$, contradicting the assumption $[k(b_0, b_1, \ldots) : k] = \infty$.

Finally, the construction is really possible, but needs inseparable extensions. Recall first the rule $(a+b)^p = a^p + b^p$ in characteristic p. This means that the pth power of a power series (or polynomial) $f = \sum a_i x^i$ is given by $f^p = \sum a_i^p x^{pi}$ (that is, it is just the power series in x^p with coefficients the pth powers of those of f). For the example, let F be a field of characteristic p such that $F^p = F$, and let $K = F(t_0, t_1, \ldots)$ and $k = K^p = F(t_0^p, t_1^p, \ldots)$, where $t_0, t_1, \ldots$ are independent transcendentals. Then the power series $f = \sum t_i x^i \in K[\![x]\!]$ does everything I said.

It is not hard to see that the normalisation $B = \widetilde{A[z]}$ is again a DVR with maximal ideal xB (see Ex. 9.1).

(2) Counterexample to chain conditions

Nagata constructs an integral domain A having maximal ideals m and n such that the residue fields are equal $A/m = A/n = k$, the localisations A_m and A_n are Noetherian, and A_m and A_n are regular local rings of different dimensions. For specific examples, see Ex. 9.3–4 or [Nagata], pp. 204–5. I use the functional notation $f \mapsto f(m) \in A/m = k$ and $f \mapsto f(n) \in A/n = k$ for the two quotient maps.

It is easy to see that if we set $S = A \setminus (m \cup n)$ and $B = S^{-1}A$ then the *semilocalisation* B has only two maximal ideals mB and nB, and is Noetherian (see Ex. 9.2). By the usual "quotient commutes with localisation" result Corollary 6.7, $B/mB = A/m = k$ and $B/nB = A/n = k$, and I continue to use the functional notation $f \mapsto f(m)$ etc. for elements of B.

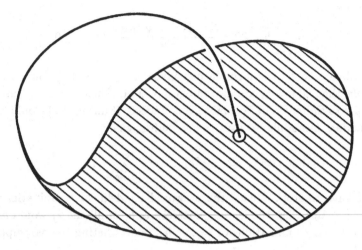

Figure 9.4. Nagata's leaf

Now "glue together" the two maximal ideals of B by setting

$$C = \{f \in B \mid f(m) = f(n)\}.$$

It is obvious that there exists an element $t \in A$ such that $t \in m$ and $t - 1 \in n$, so that $t(t-1) \in C$. It follows that $t \in B$ is integral over C, and $B = C + Ct$. The formulas here are exactly those that glue together the two points $0, 1 \in k$ of the t-line into the nodal cubic $(t^2 - t, t^3 - t^2)$. You should think of the two branches as germs of analytic manifolds crossing transversally at a point.

However, the novelty is that the two branches have different dimensions, as illustrated in Figure 9.4. Not surprisingly, questions concerning the dimension of the ring C and the codimension of its prime ideals lead to curious answers. For example, if $\dim A_m = 2$ and $\dim A_n = 3$ then the two branches of the leaf $\operatorname{Spec} A_m$ and $\operatorname{Spec} A_n$ are 2- and 3-dimensional, and contain maximal chains of irreducible subvarieties of lengths 2 and 3 respectively, giving maximal chains of irreducible subvarieties of $\operatorname{Spec} C$ of different lengths. See Ex. 9.4 for details.

(3) Counterexample to finite dimension Choose an increasing sequence of integers $0 < a_1 < \cdots < a_i < \cdots$, and let $A = k\big[\{x_{ij}\}_{i,j}\big]$ be the polynomial ring in countably many variables $\{x_{ij}\}$ indexed by $i = 1, \ldots$ and $j = 1, \ldots, a_i$. Then $P_i = (x_{i1}, \ldots, x_{ia_i})$ is a prime ideal of

A for each i, and the localisation at P_i is

$$A_{P_i} = k\big(\{x_{i'j}\}_{\forall i' \neq i}\big)\big[x_{i1}, \ldots, x_{ia_i}\big]_{(x_{i1}, \ldots, x_{ia_i})},$$

that is, a polynomial ring in a_i variables over a field localised at the origin. Set $S = A \setminus (\bigcup_{i=1}^{\infty} P_i)$. Then $S^{-1}A$ is Nagata's counterexample. Imagine the infinite dimensional vector space k^{∞} with coordinates x_{ij}, and for each i the subspace V_i of codimension a_i defined by $x_{i1} = \cdots = x_{ia_i} = 0$. Then $S^{-1}A$ is the ring of rational functions which are regular at the generic point of each V_i. Because $S^{-1}A$ has each of the A_{P_i} as a localisation, it has Krull dimension $\geq a_i$ for each i.

However, any $f \in A$ only vanishes on finitely many of the V_i, so a nonzero ideal I of $S^{-1}A$ is determined by finitely many nontrivial ideals IA_{P_i} in the Noetherian rings A_{P_i}. From this remark, the proof that $S^{-1}A$ is Noetherian is an easy exercise in the style of Chapters 3 and 6 (see Ex. 9.2).

9.5 Akizuki's example

I take up again the example $B = A[z_0, z_1, \ldots]$ and the notation from Ex. 8.5, p. 127. Assume that (a) the coefficients a_r of the power series z are units, and (b) the exponents n_r satisfy $n_r \geq 2n_{r-1} + 2$ for every $r \geq 1$, where $n_0 = 0$; in other words, the differences $m_r = n_r - n_{r-1}$ grow like $2m_r \geq n_r + 2$. (For example, the smallest possible choice is $n_r = 2(2^r - 1) = 0, 2, 6, 14, 30, \ldots$.)

Now set

$$C = A\big[t(z_0 - a_0), \{(z_i - a_i)^2\}_{i=0}^{\infty}\big] \subset B.$$

It is clear that B is integral over C, and that $\operatorname{Frac} B = \operatorname{Frac} C = K(z)$, where $K = \operatorname{Frac} A$, so that B is the normalisation of C.

Theorem $M = \big(t, t(z_0 - a_0)\big) \subset C$ *is a maximal ideal with* $C/M = k = A/(t)$, *and the localisation* C_M *has the following properties:*

(i) B_m *is integral over* C_M, *so that* $B_m = \widetilde{C_M}$ *(the normalisation).*

(ii) C_M *is a 1-dimensional Noetherian local ring.*

(iii) B_m *is not finite as a* C_M-*module.*

Proof I refer to Ex. 8.5, (1) and (2), and continue numbering formulas from there. (1) gives

$$t^{n_r} z_r = z_0 - \sum_{i=0}^{r-1} a_i t^{n_i}, \quad \text{with } \sum_{i=0}^{r-1} a_i t^{n_i} \in A. \tag{3}$$

Manipulating the identities (2–3) gives two standard tricks. First, the difference between $t(z_0 - a_0)$ and $t^{n_r+1}(z_r - a_r)$ is an element of A for any $r \geq 0$. This allows me to replace $t^{n_i+1}(z_i - a_i)$ wherever it appears by an element of A plus $t^{n_j+1}(z_j - a_j)$ with $j > i$.

Second, consider the identity

$$\begin{aligned}(z_{i-1} - a_{i-1})^2 &= (t^{m_i} z_i)^2 \\ &= t^{2m_i}\left((z_i - a_i)^2 - a_i^2\right) + 2a_i t^{2m_i} z_i.\end{aligned} \tag{4}$$

Now both terms on the right are in tC: the second, because of (3) and the assumption $2m_r \geq n_r + 2$ ((b) above). A first consequence is that *the kernel of the map $C \twoheadrightarrow k$ defined by the evaluation $t \mapsto 0$ and $z_i \mapsto a_i$ is the maximal ideal $M = (t, t(z_0 - a_0))$.*

The second trick allows me to replace $(z_{i-1} - a_{i-1})^2$ wherever it appears by

$$t^{2m_i}(z_i - a_i)^2 + \text{multiple of } t^{n_i+2}(z_i - a_i) + \text{element of } A.$$

Performing these two tricks repeatedly gives that, *for any specified $r \geq 0$ and $N > 0$, any element $f \in C$ can be written*

$$f = X + Yt^{n_r+1}(z_r - a_r) + t^N Z, \quad \text{with } X, Y \in A \text{ and } Z \in C. \tag{5}$$

Claim *For $0 \neq f \in M$, the principal ideal fC_M contains a power of t.*

Proof of Claim There exists N such that $f \notin t^N \widehat{A}$. Choose r with $n_r \geq N - 1$, and consider the expression (5). Then necessarily $X = t^n u$ with $n < N$ and u a unit of A. Dividing through by u, I assume that $X = t^n$, and

$$f = t^n(1 + t^{N-n}Z) + Yt^{n_r+1}(z_r - a_r).$$

To prove the claim, multiply f by $g = t^n(1 + t^{N-n}Z) - Yt^{n_r+1}(z_r - a_r)$:

$$fg = t^{2n}(1 + t^{N-n}Z)^2 - Y^2 t^{2n_r+2}(z_r - a_r)^2.$$

This is obviously of the form t^{2n} times an element of $C \setminus M$. Q.E.D.

I prove that the local ring C_M is Noetherian and 1-dimensional. It is clear from (5) that $C_M / t^N C_M$ is generated over $A/(t^N)$ by 1 and $t^{n_r+1} z_r$, and therefore is a Noetherian A-module. Now any nonzero ideal $I \subset C_M$ contains t^N for some N, and then the quotient ring C_M/I is also Noetherian. Therefore bigger ideals $I \subset J \subset C_M$ have the a.c.c. A nonzero prime ideal of C_M contains some t^N, and therefore also t and $t(z_0 - a_0)$, so that $\operatorname{Spec} C_M = \{0, MC_M\}$.

Under the assumption that z is transcendental, I now prove that B_m is not finite over C_M, arguing by contradiction. Since C_M is Noetherian, if B_m were finite, it would be a Noetherian C_M-module. Consider the ascending chain of submodules generated by $\{(z_i - a_i)\}_{i<r}$; for some r, I get a relation

$$z_r - a_r = \sum_{i=0}^{r-1} g_i(z_i - a_i) \quad \text{with } g_i \in C_M.$$

Writing $g_i = f_i/f_r \in C_M$ gives

$$f_r(z_r - a_r) = \sum_{i=0}^{r-1} f_i(z_i - a_i) \quad \text{with } f_0, \dots, f_r \in C \text{ and } f_r \notin M. \quad (6)$$

Now multiplying (6) by t^{n_r} and using (3) gives

$$f_r\Big(z - \sum_{j=0}^{r+1} a_j t^{n_j}\Big) = \sum_{i=0}^{r-1} f_i t^{n_r - n_i}\Big(z - \sum_{j=0}^{i+1} a_j t^{n_j}\Big). \quad (7)$$

Now all the $f_i \in C$ are polynomials in z with coefficients in A, and the left-hand side is a unit times z (because $f_r \notin M$), whereas every coefficient on the right-hand side is divisible by t. Therefore (7) is a nontrivial polynomial relation $F(z) = 0$ with coefficients in A. This contradiction completes the proof of the theorem. Q.E.D.

9.6 Scheme theory

This book has discussed the spectrum of a ring $\operatorname{Spec} A$ and its geometry as a kind of cheap formal analogue of the Nullstellensatz correspondence (see especially 5.14). This discussion can be taken as a first introduction to the theory of *affine schemes*. In this theory, the pictures of $\operatorname{Spec} A$ occuring at different places in the book become genuinely meaningful. The idea is that on top of the set $\operatorname{Spec} A$ and its Zariski topology, one must add a sheaf of rings, thought of as the functions on open sets of A. This sheaf is essentially built out of the local ring A_P over each point P, as hinted in Figure 7.2 and Ex. 7.1. Scheme theory is now more or less universally accepted as the foundation of algebraic geometry, and is an essential ingredient in many areas of number theory and commutative algebra. The student interested in pursuing this line should look at [E & H], [Sh1], [red] or [Hartshorne].

9.7 Abstract versus applied algebra

This book has enthusiastically pursued the integration of algebra and several of its substantial applications. As the student may already have noticed with distaste, this means that the book mixes up the examples and applications together with the formal axiomatic material, even when this makes the algebra apparently more complicated and harder to understand. I regard the book as a modest contribution to a currently well-established swing in fashion in world math against compartmentalised abstract algebra. There is no lack of distinguished references for this tendency, for example Atiyah's interview [A], the algebra textbooks of Michael Artin [Ar] and Kostrikin and Manin [K & M], and Shafarevich's extended essay on the meaning of algebra [Sh2].

In a nutshell, the argument is as follows: the abstract treatment of algebra, as exemplified by the work of David Hilbert, Emil Artin and Emmy Noether in the early 20th century, isolated the logical workings at the heart of very many mathematical calculations. The advantage is that you can recognise a group or a vector space at the centre of your arguments, and then you don't have to do everything all over again every time you meet one of them, because the basic material in group theory or linear algebra has already been worked out. The development of many areas of math (for example, the classification of simple Lie algebras) would be unthinkable without the abstract foundations.

The more applied view goes as follows: surely, algebra is not purely the definitions of "sets with structure" and the logical development of their properties from axioms – there would be no point in doing it unless it meant something. The ideas of algebra apply across the whole of math and science: in geometry and physics, but also in areas as diverse as pure logic, combinatorics, chemistry, other areas of abstract algebra, computer science, etc.

I discuss the background to this trend over the next few sections, restricting myself mainly to algebra as it affects undergraduate teaching in British universities in the late 20th century. The history of the Bourbaki movement, the introduction of rigorous analysis in Britain, or the rise and subsequent almost total eclipse of mathematical logic would make separate interesting stories. I would like to make it clear that there is nothing personal in my diatribe against pure algebra – indeed (as they say) some of my best friends are algebraists. My comments really refer mainly to algebra in the context of teaching, and I don't wish to imply criticism of the way other people do their research: for many research problems, abstract methods will often be the appropriate tool for the job. In Aristotle's words (quoted by Tom Körner), "it is the mark of the educated mind to use for each subject the degree of exactitude which it admits".

9.8 Sketch history

Algebra in the modern sense, with the set-theoretic foundations of math, is a 20th century phenomenon, an exact contemporary of "modernism" in art, literature and music, of the amazing superstition called psychoanalysis, and of the disastrous political ideologies of the 20th century. Abstract algebra scarcely existed in British universities until after the second world war.

It seems to me that an important background element in the history of British universities is that Britain was a world empire in the 19th century, and we were not going to take lessons from the Germans, even if they were world leaders in many areas of academic science (a clear enough perspective viewed from the U.S. or Japan). The handful of British pure mathematicians, if amiable and occasionally brilliant gentlemen, were bumbling amateurs compared to the professors of Berlin, Königsberg and Göttingen, and were critically handicapped by the lack of the Ph.D. research degree. The willful destruction of the German scientific school by the Nazis and the high profile of advanced science in the British and U.S. war effort during the second world war changed all that.

By all accounts, at least one of the formative influences on British postwar math was extremely applied: the Bletchley Park school (the "Enigma" project) used permutation groups and methods of mathematical logic in intelligence gathering and decoding, with applications to antisubmarine warfare.

The postwar introduction of set theory, abstract group theory and rigorous analysis into undergraduate courses seems to have been something of a revolution. Changes of outlook like this do not happen in isolation. Modern abstract ideas and standards of rigour were in the air from the late 19th century, and, in an unsystematic way, a few British mathematicians had been among the leaders in set theory and other modern areas. Whole new research areas such as topology were occupying some of the best minds. There must have been reaction against the old syllabus and the way it was taught. But it was probably the wartime dislocation of academic life that enabled a relatively small group of committed young mathematicians, many returning from wartime research work in defence establishments, to substantially revise the early stages of the undergraduate courses to introduce the new material.

The abstract axiomatic methods in algebra are simple and clean and powerful, and give essentially for nothing results that could previously only be obtained by complicated calculations. The idea that you can throw out all the old stuff that made up the bulk of university math teaching and replace it with more modern material that had previously been considered much too advanced has an obvious appeal. The new syllabus in algebra (and other subjects) was rapidly established as the new orthodoxy, and algebraists were very soon committed to the abstract approach.

The problems were slow to emerge. I discuss what I see as two interrelated drawbacks: the divorce of algebra from the rest of the math world, and the unsuitability of the purely abstract approach in teaching a general undergraduate audience. The first of these is purely a matter of opinion – I consider it regrettable and unhealthy that the algebra seminar seems to form a ghetto with its own internal language, attitudes, criterions for success and mechanisms for reproduction, and no visible interest in what the rest of the world is doing.

9.9 The problem of algebra in teaching

The problem in teaching occupies a quite different status, and it seems to me that it is not at all recognised: the problem is that the abstract point of view in teaching leads to isolation from the motivations and applications of the subject. For example, differential operators are typical examples of linear maps, used all over pure and applied math, but it is a safe bet that the linear algebra lecturer will not mention them: after all, logically speaking, differentiation is more complicated than an axiom about $f(\lambda u + \mu v)$, and working with infinite dimensional vector spaces would clearly needlessly disconcert the students. In the same way, if the applied people want students to study coordinate geometry in $\mathbb{R}^3$, let them set up their own course – the student who understands that the applied lecturer's $\mathbb{R}^3$ is an example of the algebra lecturer's vector spaces will be at an unexpected advantage. Similar examples occur at every point of contact between algebra and other subjects; under the system of abstract axioms, the algebraist is *never* going to take responsibility for relating to the applications of his subject outside algebra.

No subject has suffered as badly from the insistence on the abstract treatment as group theory. When I was a first year undergraduate in Cambridge in 1966, it had been more or less settled, presumably after some debate, that the Sylow theorems for finite groups were too hard for Algebra IA; since then, the notion of quotient group, and subsequently the definitions of conjugacy and normal subgroup have been squeezed out as too difficult for the first year. Thus our algebraists have cut out most of the course, but stick to the dogma that a group is a set with a binary operation satisfying various axioms. Groups can be taught as symmetry groups (geometric transformation groups), and the abstract definition of group held back until the student knows enough examples and methods of calculation to motivate all the definitions, and to see the point of isomorphism of groups.

The schizophrenia between abstract groups and transformation groups comes to the surface in some amusing quirks – for example, the textbooks that define an "abstract group of operators", or the students (year after year) who insist that the binary operation $G \times G \to G$ on a group should satisfy closure under $(g_1, g_2) \mapsto g_1 g_2$ as one of the group axioms.

It seems to me that the abstract approach has weaknesses even within the framework of abstract algebra. In recent years, the Warwick 3rd year has featured a course on Lie algebras. I've no doubt that the course is extremely well given, but it's still possible to find students who get good grades, and know all the bookwork in the course, but who still don't know that $n \times n$ matrixes over $\mathbb{R}$ with bracket $[A, B] = AB - BA$ is an example of a Lie algebra, and $\mathbb{R}^n$ with $Av = $ matrix times a vector an example of a representation or a module. The student who knows just this one example can make good sense of the entire course. Of course, given a chance, any self-respecting geometer, applied mathematician or physicist would insist on muddling things up by differentiating the group law at the origin, and explaining what happens to the associative law, etc. Is it conceivable that there are people about who introduce the Jacobi identity as a bald axiom?

9.10 How the book came to be written

I have lectured this material five or six times since first coming to Warwick in 1978, and my notes have been used more recently for 3rd year reading courses, and as part of the requirement for an M.Sc. reading course. At first I had great difficulty sorting out what I wanted to say in commutative algebra from the more geometric material that went into [UAG]. At the same time, several of my colleagues have told me that it would be a waste of time just rewriting [A & M], so I've had to make an effort to distinguish my presentation from theirs.

The title of the book and the pictures are intended to be provocative.

Acknowledgements

Much of the material is cribbed consciously or unconsciously from the wonderful books of Atiyah and Macdonald [A & M] and Matsumura [M]; in particular I have used as exercises several really pretty ideas from [M]. Colin McLarty made a whole series of useful comments on the text and the exercises, and urged me to fill out the motivation at a number of points. Jack Button and Alan Robinson helped with references to number theory; I would also like to thank Gavin Brown and many Warwick students for questions and corrections, and Charudatta Hajarnavis, David Epstein and John Moody for remarks used in the text. I am grateful to David Fowler and Jeremy Gray for offering advice (not necessarily followed) on my historical digression.

Jokes

I have written myself innumerable little "to do" notes in the course of preparing the final draft of this book. I hope that most have been tidied up somehow or other, but one in particular stubbornly resists all attempts, and I leave it as a challenge for future generations:

> Think of some jokes? Curiously, it is much harder to
> make jokes in an algebra course than in geometry.

Exercises to Chapter 9

9.1   Let $B = \widetilde{A[z]}$ be as in 9.4, (1). By using the fact that B is integral over A prove that every nonzero maximal ideal of B lies over Ax. Deduce that B is a local ring. Now the usual argument of Proposition 8.3 gives that B is a DVR with local parameter x.

9.2   Prove that the ring $S^{-1}A$ of 9.4, (3) is Noetherian. [Hint: prove that an ideal $J \subset S^{-1}A$ is uniquely determined by the ideals

$JA_{P_i} \subset A_{P_i}$ for every i by applying the arguments of 6.3 to the localisations $S^{-1}A \to A_{P_i}$. Obviously, any nonzero polynomial $f \in A$ is only contained in finitely many of the P_i; deduce that if I is a nonzero ideal of $S^{-1}A$ then $IA_{P_i} = A_{P_i}$ for all $i > n_0$. Therefore, an ideal J containing I is determined by the finitely many ideals JA_{P_i} of Noetherian rings A_{P_i}. Conclude as usual by the a.c.c.]

9.3 Let $z = \sum_{i=0}^{\infty} a_i x^i \in k[\![x]\!]$ be a formal power that is transcendental over $k[x]$. Consider the ring $A = k[x, z_0, z_1, \ldots]$ generated by variables x, z_i subject to the following relations:

$$z_i = x z_{i+1} + a_i \quad \text{for } i = 0, 1, \ldots.$$

(a) Construct an isomorphism of A with a subring of $k[\![x]\!]$ sending z_0 to z.

(b) Prove that the kernel of the homomorphism $A \to k$ defined by $x \mapsto 0$ and $z_i \mapsto a_i$ is the principal ideal $m = (x)$.

(c) Prove that the localisation A_m is a DVR. [Hint: Ex. 8.4 was set up for this purpose.] In other words, the z_i are formal power series in x, so they behave like function of x near $x = 0$, and $\operatorname{Spec} A_m$ is a 1-dimensional nonsingular analytic arc.

(d) Prove that $A[1/x] = k[x, 1/x, z_0]$.

(e) Fix any $b \in k$, and prove that $n = (x - 1, z_0 - b)$ is the kernel of the homomorphism $A \to k$ defined by $x \mapsto 1$ and $z_j \mapsto b - \sum_0^{j-1} a_i$, and is hence a maximal ideal.

(f) Prove that $A_n = k[x, z]_{(x-1, z-b)}$ is the local ring of the x, z-plane at the point $x = 1, z = b$, and is therefore a Noetherian ring. [Hint: use the previous two items and the assumption that x, z are algebraically independent.] For the reader who knows the definition, A_n is obviously a 2-dimensional regular local ring.

(g) Let $S = A \setminus (m \cup n)$ and set $B = S^{-1}A$, the *semilocalisation* of A at m, n. Check that B is a Noetherian ring (the argument is similar to Ex. 9.2) with two maximal ideals B_m and B_n.

(h) Define C as in 9.4, (2), and verify that C is Noetherian. [Hint: prove that every ideal $I \subsetneq C$ is an ideal of B, then use the a.c.c.]

(i) Prove that the prime ideals of C are in one-to-one correspondence with the prime ideals of A_n. [Hint: A_m has
only two prime ideals, 0 and $mA_m = (x)$.]

9.4   Adding a variable y to the construction of the previous exercise
gives a ring C that is not *catenary*: it has maximal chains of
prime ideals of different lengths, so that its Krull dimension is
ambiguous.

Let $z \in k[\![x]\!]$ be as in Ex. 9.3, and consider the ring $A = k[x, y, z_0, z_1, \dots]$ generated by variables x, y and z_i subject to
the same relations. As before, prove that $m = (x, y)$ and
$n = (x - 1, z_0 - b, y)$ are maximal ideals with residue field k,
and the localisations A_m and A_n are regular local rings with
regular parameters (x, y) and $(x - 1, z_0 - b, y)$. Construct the
semilocalisation $B = S^{-1}A$ at n and m, and the subring $C \subset B$
obtained by "glueing together" the two maximal ideals of B by
the isomorphism $B/m = B/n = k$. Then $C, m_C = mA_m \cap nA_n$
is a local ring.

By intersecting C with maximal chains of prime ideal $0 \subset P_1 \subset mA_m$ in A_m and $0 \subset Q_1 \subset Q_2 \subset nA_n$ in A_n, prove that
it has maximal chains of prime ideals of length 2 and 3. For
example,

$$0 \subset (x) \subset (x, y) \quad \text{in } A_m$$

and

$$0 \subset (x - 1) \subset (x - 1, z_0 - b) \subset (x - 1, z_0 - b, y) \quad \text{in } A_n.$$

9.5   In Akizuki's example 9.5, use (4) to prove that $M^2 = tM$.

9.6   Prove that $t^{n_r}(z_r - a_r) \notin C$ for any $r \geq 0$. [Hint: following
the method of 9.5, use $t^{n_r}(z_r - a_r) \in C$ to derive an algebraic
dependence relation for z over A.]

Bibliography

[Akizuki] Y. Akizuki, Einige Bemerkungen über primäre Integritätsbereiche mit Teilerkettensatz, Proc. Phys.–Math. Soc. Japan, **17** (1935), pp. 327–336

[A] An interview with Michael Atiyah, *Math. Intelligencer* **6**:1 (1984), 9–19

[Ar] M. Artin, *Algebra*, Prentice–Hall, Englewood Cliffs, NJ, 1991

[A & M] M. F. Atiyah and I. Macdonald, *Introduction to commutative algebra*, Addison–Wesley, Reading, Mass., 1969

[B & Sh] Z. I. Borevich and I. R. Shafarevich, *Number theory*, Nauka, Moscow, 1972; English translation: Academic Press, New York, 1976

[C] P. M. Cohn, Algebra, Wiley, London, 1989

[E & H] D. Eisenbud and J. Harris, *Schemes: The language of modern algebraic geometry*, Wadsworth and Brooks/Cole, Pacific Grove, CA, 1992

[EGA] A. Grothendieck and J. Dieudonné, *Eléments de géométrie algébrique*, IV, (Seconde Partie), Publications Math. IHES **24**, 1965

[F] W. Fulton, *Algebraic curves*, Addison–Wesley, Redwood City, CA, 1989

[H & H] B. Hartley and T. O. Hawkes, *Rings, modules and linear algebra*, Chapman and Hall, London, 1970

[Hartshorne] R. Hartshorne, Algebraic geometry, Springer, Berlin, Heidelberg, New York, 1977

[K & M] A. I. Kostrikin and Yu. I. Manin, *Linear algebra and geometry*, Moscow Univ. Press, Moscow, 1980; English translation: Gordon and Breach, New York, London, 1989

[M] H. Matsumura, *Commutative ring theory*, Cambridge, 1985

[red] D. Mumford, *The red book of varieties and schemes* (it's actually yellow), Springer LNM 1358, Springer, Berlin, Heidelberg, New York, 1988

[Nagata] M. Nagata, Local rings, John Wiley Interscience, New York, London, 1962

[N] T. Nagell, *Introduction to number theory*, Chelsea, New York, 1964

[UAG] M. Reid, *Undergraduate algebraic geometry*, C.U.P., Cambridge, 1988

[Sh1] I. R. Shafarevich, *Basic algebraic geometry* (2 vols.), Springer, Berlin, Heidelberg, New York, 1994

[Sh2] I. R. Shafarevich, *Basic notions of algebra, Springer Encycl. of Math. Sciences* **11**, Springer, Berlin, Heidelberg, New York, 1990

[W] B. L. van der Waerden, *Algebra* (2 vols), Springer, Berlin, Heidelberg, New York, 1991

Index